ALSO BY ROBERT M. SAPOLSKY

Why Zebras Don't Get Ulcers:
A Guide to Stress, Stress-Related Diseases, and Coping

Stress, the Aging Brain,
and the Mechanisms of Neuron Death

THE
TROUBLE
WITH
TESTOSTERONE

and Other Essays on the Biology
of the Human Predicament

∞

ROBERT M. SAPOLSKY

A TOUCHSTONE BOOK
PUBLISHED BY SIMON & SCHUSTER

TOUCHSTONE
Rockefeller Center
1230 Avenue of the Americas
New York, NY 10020

First Touchstone Edition 1998

TOUCHSTONE and colophon are registered trademarks of Simon & Schuster Inc.

DESIGNED BY ERICH HOBBING

Set in Garamond No. 3

Manufactured in the United States of America

9   10

The Library of Congress has catalogued the Scribner edition as follows:
Sapolsky, Robert M.
The trouble with testosterone : and other essays on the
biology of the human predicament / Robert M. Sapolsky.
p. cm.
Includes bibliographical references.
(alk. paper)
1. Psychobiology. I. Title.
QP360.S268 1997
610'.1—dc21   96-52357
CIP

ISBN 0-684-83409-X
ISBN 0-684-83891-5 (Pbk)

"How Big Is Yours?"; "The Young and the Reckless"; "Junk Food Monkeys"; "The Graying of the Troop"; and "The Dissolution of Ego Boundaries and the Fit of My Father's Shirt" were previously published in *Discover*. "Primate Peekaboo"; "Measures of Life"; "The Solace of Patterns"; "Poverty's Remains"; "The Burden of Being Burden-Free"; and "Curious George's Pharmacy" were previously published in *The Sciences*.

Contents

⧬

For my mother, who,
a long time ago,
read to me

THE
TROUBLE
WITH
TESTOSTERONE

Introduction

We all have encountered Reinhold Niebuhr's serenity prayer at some point:

> *God, grant me the serenity to accept the things I cannot change, courage to change the things I can, and wisdom to know the difference.*

Behavioral biology is often the scientific pursuit of that prayer. Which of our less commendable ways of behaving, asks the behavioral biologist, can we hope to change (and how) and which are we stuck with? Asked in a harsher way—as our society so often poses these questions of nature and nurture—for which of our failings should we be held responsible? Did Charles Whitman open fire from the University of Texas observation tower and kill eighteen people because of his brain tumor? Did Richard Speck murder eight nurses because of his alleged extra Y chromosome? Did Dan White kill San Francisco mayor George Moscone and city supervisor Harvey Milk because of his "diminished capacity": attributable, in part, his lawyers claimed, to a junk food addiction? Did John Hinckley shoot President Reagan because of insanity? Or were they all just rotten characters? What about the spouse sunk in depression? Is a neurochemical imbalance to blame, or is the person just indulging in a profound sulk? Is the floundering schoolchild limited by a learning disability, or plain lazy?

In the most narrow sense, behavioral biologists seek to answer questions such as these by exploring the interface between our minds and our bodies. How is it that you can think a thought, have a memory or a surge of emotion—products of our minds—and, as a result, alter the activities of virtually every cell in the body? And, in turn, what are the mechanisms by which events in our bodies—changes in hormonal status, nutrition, health—can change our thoughts and feelings? Answering questions such as these begins to answer the broadest questions of all—what is the biology of what makes us who we are, what is the biology of our individuality, our limits and potentials?

This is frightening ground to tread, partly because of the complexity of the questions asked. It's easier to determine how birds navigate while migrating or how muscle fibers contract than to answer a question like "Is there a genetic basis to criminality?" Scarier still are the abuses to which this work is subject. It's difficult to become an ideologue about bird migration or muscle physiology, but behavioral biology is a magnet for those with an ax to grind. Conscientious scientists fear that a minute observation, tenuously offered, might be seized upon by someone eager to lend scientific authority to claims like "I'm not responsible for my problems," or worse, "I don't have to help you in combating your problems, because they are incurable." At one extreme can be a wasted life, when prejudice dictates that there is a limit that does not exist. Witness those who have been discriminated against by race, ethnicity, or gender because they were believed to be biologically—and therefore, in this view, irredeemably—inferior. At the other extreme, the specter of blame can haunt the blameless, when ignorance leads to failure to recognize a real biological constraint that exists. In that case, witness the generations of dyslexics wrongfully condemned as stupid.

No matter how obscure the subfield of science, there is bound to be some crazed egghead out there who finds it fascinating. When it comes to the biology of our individuality, issues are raised that should be fascinating to each of us for the simple reason that we are all asked to function as behavioral biologists on some occasion. Most importantly, those occasions matter: We serve on juries and decide the culpability of someone regarding something awful that they have done. We are asked to vote on referendums about expenditures of public funds to try to fix some social ill, and we have to decide if we think it can be fixed. We see someone stymied in their learning, claiming to be at their limits, and we must decide whether pushing them harder represents inspiration or cruelty. And perhaps we will have to watch a loved one decline with some terrible disease, watch their personality be transformed, and have to learn that this is due to the illness and not to them.

Insofar as we are forced to be practicing behavioral biologists, we might as well be competent at it. This collection of essays is meant to help a bit in that direction in that it offers a tour of the field (albeit a highly unsystematic one, which is to say that it covers a hodgepodge of topics that I'm particularly fascinated by). Broadly, the essays fall into three categories. One group presents some of the latest breakthroughs in psychiatry, neuroscience, and endocrinology. One might assume that much of these findings will be about big, messy problems in abnormal human behavior, about mental illness, uncontrollable violence—the arena of "them and their diseases." As will be seen, some of the most provocative findings in the field are a lot closer to home, are about far more subtle issues—why we differ in our sexual orientation, in our desire for novelty, in styles of thinking and feeling; this is not about them and their diseases, but about the biology of the quirks and idiosyncrasies of our everyday behaviors. It is my experience from lecturing about these topics that people get fairly disturbed by the implications of some of these findings, as we wind up seeming to have a lot less volition in our behaviors than most like. The extreme example of this is "Circling the

Blanket for God," the final essay in the collection, in which I consider some of the neuropsychiatric roots of religious belief.

Another set of essays explores many of these same issues from the perspective of evolutionary biology and animal behavior. Initially, many of these pieces will seem to be about how some of our close relatives—nonhuman primates, for example—turn out to be vastly more subtle and complex in their behaviors and emotions than one would ever have guessed from watching *Wild Kingdom*. What often only sinks in next is not only how subtle and complex they are but how familiar, reinforcing the lesson that we humans are just another primate species: a terribly neurotic, screwed-up, overly self-conscious one with some fancy thumbs, but still just another primate. As but one example of this style, "Primate Peekaboo," written a year ago when, to my embarrassment, I was devoting about ten hours a day to thinking about the O.J. trial, considers the intense voyeurism that is common to all of us primates.

Finally, another group of essays considers some of the political or social implications of findings in these areas. Some of these are historical, reviewing some of the disastrous dead-ends in behavioral biology, some of which are the end-product of well-intentioned mistakes, some the end-product of anything but good intentions. For example, in "Poverty's Remains," I recall the history of an imaginary disease that was invented around the turn of the century because scientists did not yet know anything about how stress affects the body, and before it was over with, thousands of people were killed by the consequences of medical belief in this erroneous discovery. And some of these pieces are meant to alert us to some of the dangers to come. In "How Big Is Yours?" for example, I discuss some recent and controversial evidence that the size of a certain sliver of the brain has something to do with a man's sexual orientation and then raise the question—what happens when brain imaging techniques get to the point, as they soon will, where a ten-year-old can be informed as to the size of that silver in his own brain?

Thus, the ground that will be covered. A final note: I believe that everyone can benefit from learning some science these days, regardless of how efficient a job those teachers did in junior high school convincing us that we hated the subject. When science works right, it is an amazing thing to behold—it provides us with some of the most elegant, stimulating puzzles that life has to offer. It throws some of the most provocative ideas into our arenas of moral debate. And occasionally, it even improves our lives. The subjects contained in here, I feel, have the abundant potential to accomplish all of that; thus, I have made an effort to write these essays so that they will be accessible to anyone, even the card-carrying science-phobic. I think I can guarantee that the facts in here will be relatively simple, and I know I can guarantee that their implications will not be.

Acknowledgments

Some of the topics covered in these essays are related to my area of research, and I am supposed to know something about them. In many cases, though, these are about subjects that are not in my area of expertise (which, as will be seen, hasn't prevented me from developing some addled opinions about them). In these cases, I have been very dependent on the generosity and clarity of some of my colleagues in discussing their work and ideas. I thank Jeanne Altmann, Jay Belsky, Laurel Brown, Jonathan Cobb, Jared Diamond, Bill Durham, Laurence Frank, James Gross, Ben Hart, Charles Nemeroff, Craig Packer, Edward Paul, Larry Squire, Michele Surbey, Andrew Tomarken, the late Amos Tversky, and Richard Wrangham for their assistance; any factual errors are of my own doing.

I am particularly grateful to the research assistants who helped me with many of these essays—Steve Balt, Roger Chan, Michelle Pearl, Dave Richardson, Serena Spudich, and Paul Stasi. They have been heroic in finding the most obscure of facts in the farthest corners of archival libraries, surfing the Internet for answers to trivia questions, and laboring at odd hours of the night to call experts on the opposite side of the globe. I also thank Helen and Peter Bing, a visionary couple who have made grants available to various Stanford professors for forays into unorthodox teaching initiatives. Most of these essays started off as lectures, and the Bings made the labors of these assistants possible.

I also thank the teachers who got me into science and, just

as much, made me excited about wanting to communicate about science to nonexperts: the late Howard Klar, Howard Eichenbaum, Melvin Konner, Bruce McEwen, Lewis Krey, Paul Plotsky, and Wylie Vale.

Many of these pieces were originally published in *Discover,* or in *The Sciences,* and the writing was often greatly improved by my editors there. It has been a great pleasure working with them, and I thank them all—Burkhard Bilger, Peter Brown, Alan Burdick, Robert Coontz, Patricia Gadsby, Paul Hoffman, Richard Jerome, Polly Shulman, and Marc Zabludoff. Those readers who are familiar with *The Sciences* will know of the superb artwork that graces its pages; that is the work of Liz Meryman, and I thank her for consenting to read these pieces and give advice on accompanying artwork.

Walter Bode rather indirectly made it possible for these essays to see the light of day, and I thank him for that. Liz Ziemska, my agent, far more directly made it possible, and my great thanks to her as well. Hamilton Cain and Jennifer Chen have helped every step of the way at Scribner, and it has been a great pleasure to work with them.

Finally, my gratitude to my wife and best friend, Lisa, for endless numbers of reasons. However, there are special grounds for gratitude with respect to this book—during vacations, during times when she has been struggling with some deadline, during times when she has just been trying to get some sleep; during good times and bad, through thick and thin, in sickness and in health, she has put up with my excited ranting and raving about whichever of these essays I've been working on at the time.

How Big Is Yours?

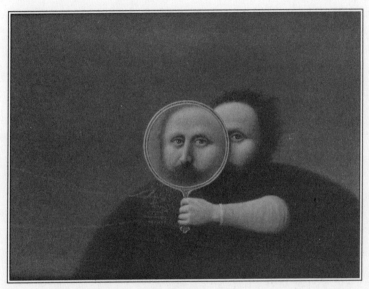

Alfredo Castañeda, When the Mirror Dreams with Another Image, *1988;
courtesy Mary-Anne Martin/Fine Art, New York*

During my graduate school days in New York City I lived along the East River, and at times when I felt like indulging a simultaneous sense of adventure and melancholy I would visit Roosevelt Island. The island was a sliver of land in the river, two and a half miles long, accessible from Manhattan by a pleasing aerial tramway. Today most of Roosevelt Island is filled with high-rise apartment buildings. But in earlier times it was the dumping ground for various incorrigible or unmanageable members of society. At the very tip of the island are some remnants of those times—the rubble of a mental asylum abandoned in the first half of this century.

A decade ago it was still possible to climb around in those ruins. You could shin up the banister of a staircase whose steps had long since decayed away, push open creaking metal doors half off their hinges, and enter a room without a roof. You could then tiptoe through a third-floor hallway about to give way and hurtle you through the splinters into the basement (and, you were sure, into a nest of rats the size of pit bulls).

It was impossible to inch through the debris without being moved by the events that must have taken place in this ghost of Bedlam. There were doors marked INSULIN SHOCK ROOM, rusted gurneys with restraining straps teetering halfway through holes in the floor, and bloodstains on the walls. Even on a warm autumn day with the sun shining on the roofless building, the whole place still felt dank and shadowed, the walls humid with the screams of misery and sadness.

Contemplating the treatment of an insane person from a century ago is something of a Rorschach test for us. Do we focus on the vast progress that has been made in psychiatry? Or do we see no difference at all from our own miserably inadequate treatment of the mentally ill?

Some things remain depressingly the same across the centuries: In so many times and places, the mentally ill give the rest of us the willies, and they are carefully isolated and ostracized. Yet many other things have changed. When we discuss treatments now, we think of drugs to manipulate brain chemicals such as neurotransmitters, while in earlier times it was lobotomies and insulin-induced comas, and still earlier, restraint and ice baths. Our notions of causes have changed as well. Now we discuss receptor regulation and genes, while earlier we would have blamed mothers sending conflicting signals of love and hate to impressionable young children.

What has changed most palpably, however, is our attitude toward abnormal behavior. We have become far more subtle when we consider the thorny issues of blame. Centuries ago epileptics were persecuted for their presumed bewitchment. We no longer do that, nor would any rational person prosecute an epileptic for assault and battery should that epileptic injure someone while flailing during a seizure. We have been trained to have a strikingly compassionate thought that is one of the triumphs of our century: "It's not him. It's his disease." We have been taught to draw a line between the essence of a person and the neuropsychiatric disorder that distorts and constrains that essence.

We are very good at drawing that line in rejecting the idea that an epileptic is violent because his arms move uncontrollably during a seizure. But we are not particularly good at drawing the line between a person and his disease in many other realms. Witness, for example, the Neanderthal bellowing in so many editorials about how John Hinckley was "getting away with it" when he was hospitalized as a schizophrenic rather than being jailed after shooting Reagan. Or contemplate the number of teachers and parents who are not very good at drawing the line between the essence of who a child is and the learning disabilities that impinge on that essence—and who instead let words like "lazy" or "stupid" creep in.

If many of us are not very good at drawing that line now, that problem is going to get worse. Some astonishing new trends in neuropsychiatry and behavioral biology indicate that the line will shift in directions we never would have guessed. This shift affects much more than our understanding of the biological imperatives that drive a small group of us to monstrous behavior. It also affects how we view the quirks and idiosyncrasies that make each of us a healthy individual.

To me, one of the most intriguing changes has occurred in the way we see "schizotypal" individuals. A few decades ago a team headed by psychiatrist Seymour Kety of Massachusetts General Hospital initiated studies that demonstrated a genetic component to the disordered jumble of thoughts known as schizophrenia. The scientists examined adoption records meticulously maintained in Denmark, reviewing the cases of children adopted from their biological parents very early in life. If a child of a schizophrenic parent was adopted by healthy parents, Kety wanted to know, was the child at greater than average risk for schizophrenia? Conversely, did any child of healthy biological parents raised in a household with a schizophrenic adoptive parent have an increased risk for the illness?

Kety's work showed that genetics does in fact increase the likelihood of the disorder. But to get that answer, doctors had to conduct intensive psychiatric interviews with the various biological and adoptive parents. This involved thousands of people and years of work. No one had ever studied the relatives of schizophrenics in such numbers before. And along the way someone noticed something: a lot of these folks were quirky. These relatives were not themselves schizophrenic—just a bit socially detached and with a train of thought that was sometimes a little hard to follow when they spoke. It was something mild, and not the sort of thing you'd note in talking to the family members of a few schizophrenics, but it suddenly stuck out when you dealt with thousands of them. They believed in strange things and were often overly concerned with magical or fantasy thinking. Nothing certifiably crazy—maybe a heavy interest in science fiction and fantasy, or a firm belief in some New Age mumbo jumbo or astrology, or perhaps a very literal, fundamentalist belief in biblical miracles. None of these are illnesses. Many adults attend Star Trek conventions, presidents' wives consult astrologers and are still taken seriously by the fashion industry, and others believe that the earth really was created in seven days. But today psychiatrists call the collection of traits seen by Kety "schizotypal personality disorder," especially the emphasis on magical thinking and the loosely connected thoughts. Apparently, if you have a certain genetic makeup, you're predisposed to schizophrenia. Have a milder version of this genetic makeup, and you may be predisposed to placing a strong faith in magical ideas that are not particularly based on fact. Is there a gene for believing in the Force and Obi-Wan Kenobi? Certainly not, but perhaps there's something closer to it than we ever would have imagined.

Behavioral biology is also revealing the workings of our normal inhibitions. Over the course of an average day there must be a dozen times in which you have a thought—lustful or angry or petulant or self-pitying—that you would *never ever* say. Damage a certain part of your brain's frontal cortex and you now say those things; the frontal cortex is the closest thing we have, neuroanatomically, to a superego. Phineas Gage, a nineteenth-century railroad worker, wound up a celebrated neurological patient and fairground exhibit after his left frontal cortex was destroyed in a freak accident. He was transformed from a taciturn man to a pugnacious loudmouth who told everyone just what he thought. "Frontal disinhibition," involving aggressiveness, inappropriate frankness, and hypersexuality, is also often seen in individuals who have sustained stroke damage to that part of the brain. Remarkably, the same appears to happen in Huntington's disease, a rare congenital neurological disorder. Scientists have long thought of the disease as a movement disorder—around age forty to forty-five, patients begin to demonstrate uncontrolled swinging of limbs as an inhibitory motor pathway in the brain degenerates. With time, the movement becomes all-encompassing, constant whole body writhing that incapacitates the person. A lesser-known feature of the disease is a social disinhibition, one that often even precedes the motoric aspects, and in recent years it has been shown that Huntington's individuals also have damage to their frontal cortex.

Some neuroscientists even use the word "frontal" in a sardonic sense: A terrified student gives a quavering lecture to his elders, and some insensitive big shot rises and savages the kid over some minor point, taking the opportunity to toot his own horn while he's at it. "Christ," someone will mutter in the back of the lecture hall, "he's getting more frontal all the time."

Blow away that part of the brain and you can still remember the name of your kindergarten teacher, still do a polka, still feel what all of us feel. You just let other people know about it far more often than do most of us. Is it absurd to hypothesize that there is something a little bit wrong with the frontal cortex of the insensitive big shot in the lecture hall?

Another version of neuropsychiatric disinhibition is seen in Tourette's syndrome, once a diagnostic backwater but fast threatening to become a fad. Tourette's patients are famed for their scatology, their uncontrolled cursing. But this doesn't even begin to scratch the surface. Tourette's patients do indeed curse, but they also emit a stream of animal sounds—yips and barks and growls—along with facial tics, and violent or lewd body gestures. These are the first genetic and neurochemical hints as to what the disease is about, but it remains, for the most part, a mystery. What is striking, though, is how it differs from the disinhibition of a frontal patient. A frontal individual does or says what the rest of us think about but would never let out of our well-restrained minds. Tourette's patients do not wish to bark like a dog or grab repeatedly at their crotch—these are simply emotive twitches, uncontrolled outbursts that are randomly tossed on top of the person struggling to maintain continuity. Like hiccups of the id.

Thus, a variety of these neuropsychiatric disorders result in marked and puzzling disinhibition. Some epileptics undergo a personality shift in the opposite direction. Roughly defined, an epileptic seizure is an abnormal electrical discharge in the brain. Neurologists have known for a long time that just before the onset of a seizure there will often be a strange sensation, or "aura," and the location of the seizure in the brain can influence the type of aura—for example, epileptics will typically have a sensory aura, perhaps imagining a particular smell. The existence of auras demonstrates the not very surprising fact that sudden bursts of electrical activity in different parts of the brain will influence thought and sensation. Now neurologists are coming to recognize that different types of epilepsy also shape personalities, influencing the person all the time, not merely seconds before a seizure.

People with a type of temporal lobe epilepsy, for example, tend to be extraordinarily serious, humorless, and rigid in their ways. They tend to be phobic about doing new things, and instead perseverate on old behaviors and tastes, tending to walk to work the same way each day, usually wearing the same types of clothes, ordering the same meal in restaurants. Similarly, they rely upon a very small circle of friends, showing what neuropsychologists pungently call a "viscous" or "sticky" personality. Such people also tend to have an intense interest in religion or philosophy. And, most oddly, they not only think obsessively about their problems, they write about them—endlessly. Temporal lobe epileptics are renowned among neurologists for this "hypergraphia." In a typical scenario, someone first seeing a new neurologist will present the doctor with a carefully handwritten eighty-page diary, insisting that reading it will give the doctor vital insight into the patient. At the next visit the epileptic will return with a new, fifty-page addendum. One might speculate that having a serious neurological illness like epilepsy would make anyone serious and cause people to focus on the philosophical things in life, narrow their horizons, and rely on comfortable, familiar patterns. But this personality change does not arise from other types of epilepsy of an equally serious nature, is not a function of the frequency or severity of seizures or the magnitude with which it disrupts a person's life. Instead, have an uncontrolled and rhythmic outburst of electrical activity in the temporal lobe every now and then and, the rest of the time, you get very interested in philosophy and always order the same meal in a restaurant.

There is another version of a constrained life that is being defined biologically. At some time each of us has, to our irritation, left on a trip and felt such nagging doubt as to whether we locked the door that we returned home to check. Or after dropping a letter into a mailbox, we have peeked in a second or third time just to make sure it went down. Or, during a tough, anxious period in our lives, we find ourselves unable to concentrate because some ridiculous television jingle keeps running through our heads. This is normal and common. But among people with obsessive-compulsive disorder, these thoughts dominate and ruin their lives. They miss vacations because they return home repeatedly to check if the oven was turned off. They lose their jobs because they are late each day, spending hours each morning washing their hands. They torture themselves by obsessively counting numbers in their heads. For most of us, little rituals of thought or behavior can calm us and provide structure at an anxious time. For someone with obsessive-compulsive disorder—now thought to be caused by an imbalance of brain chemicals, possibly serotonin and dopamine—there are no limits, and the person becomes a creature of these rituals.

What does this tour of neuropsychiatric oddities mean? We are learning to draw that line in new places. Most of these disorders did not exist a few decades ago; we did not even have names for how biology could occasionally destroy the life of an individual. Now we have those names. We are beginning to learn what certain parts of the brain, what specific genes, or what our early development has to do with these tragedies. In the process we are extending our definition of illness. For some time we have generally accepted that people who rave and gibber are ill, that they cannot control these things, are made miserable by them, and deserve care, protection, and forgiveness. Slowly we are coming to recognize that you can also be made miserable by a ceaseless march of number counting in your head, or by paralyzing fears of anything new, and that these too can be uncontrollable illnesses that demand understanding and treatment.

This field continues to move forward, and we might even be able to cure some of these maladies. Another form of progress will be the recognition of increasing numbers of these disorders, the coining of more names to describe our behavioral oddities. What will happen when, eventually, we have a few of these labels?

I recognize facets of myself in these pages. At times when I am overworked and anxious, I develop a facial tic and I count stairs as I climb them. I usually wear flannel shirts. In Chinese restaurants I always order broccoli with garlic sauce. Invariably I think, "I'll get broccoli and garlic sauce," then I think, "Nah, order something different," then I think, "Well, I enjoyed broccoli last time, why get something different?" and then I think, "Careful, I'm becoming a perseverating drudge," and then the waiter is standing there and I become flustered and order broccoli with garlic sauce.

I do not have temporal lobe epilepsy, obsessive-compulsive disorder, or any of the other problems I have discussed. Yet it is reasonable to assume that there is some sort of continuum of underlying biology here—whatever it is about the temporal lobe of some epileptics that makes them perseverate may share some similarity with my own temporal lobe, at least when it is menaced with options like Buddha's Delight or General Po's Szechwan Chicken. Perhaps whatever neurochemical abnormality causes a schizophrenic to believe that voices are proclaiming her the empress of California is the same abnormality that, in a milder form, leads a schizotypal person to believe in mental telepathy. In an even milder form it may allow the rest of us to pass a few minutes daydreaming that we are close friends with some appealing movie character.

What if eventually we come to understand the genetics, the neurochemistry, and the hormonal bases of clothing preference, of who votes Democratic, of religiosity, or of why some worry too much about money and others too little? Some of these irritating traits that are, at worst, character weaknesses, but nothing more pathological. Slowly we will be leaving the realm of *them* and their disorders. We will be defining instead a biology of *us* and our strengths and weaknesses, of our potentials and constraints.

In 1992, the newspapers were teeming with stories about one such advance. For years scientists have searched for differences between heterosexual and homosexual men, and nothing very consistent had ever emerged. But in August 1992, the prestigious journal *Science* published a paper by neurobiologist Simon Levay demonstrating just such a difference. And it's a whopper of an interesting one. It concerns the hypothalamus, a part of the brain central to sexual behavior. The size of one subregion at the front of the hypothalamus, known by the not terribly titillating title of the third interstitial nucleus, differs by sex; males have a larger one than do females. Levay reported that homosexual men have smaller nuclei than do heterosexual men—as small, in fact, as those found in women.*

For some homophobes this is a bellwether observation: "You see, there is something wrong with their brains." For some gays it is an affirmation: "You see, I've always told you I just felt gay. This is what I was meant to be." And for those concerned with social policy, fearing that a gay teacher will make children gay is as preposterous as a blue-eyed teacher influencing children's eye color. Predictably, the scientific jury is still out: while Levay is a superb neuroscientist, his sample size was small, and the brain tissue he examined came from AIDS patients, so the disease might have altered it. Also, Levay doesn't know if the small size is the cause or the result of sexual orientation.

But suppose his finding turns out to be accurate. And suppose a small interstitial nucleus in a male turns out to be more a cause than a consequence of homosexuality. What will happen when brain imaging techniques improve to the point, as they inevitably will, where we can measure the size of this brain structure in a person? Being gay is not a disease. (If you don't

*This has given rise to a T-shirt that I saw being worn in the Castro District of San Francisco: "The only thing small about me is my third interstitial nucleus." It's great to see when neuroanotomy begins to influence fashion.

believe me, just ask the American Psychiatric Association. Being gay *used* to be a mental illness until the APA, in a spasm of political correctness and enlightenment, changed its mind and struck homosexuality from its bible, the *Diagnostic and Statistical Manual of Mental Disorders*. Overnight, millions of people had one less disease, which is a pretty impressive outcome of a committee meeting.) Do we inform adolescents of their nucleus size when they have not yet become sexually active and haven't expressed a sexual preference? What do we do with an openly and happily gay or straight adult whose nucleus is the "wrong" kind? What will we make of nuclei of intermediate sizes? And will the Food and Drug Administration move to squelch the predictable festering of quacks flogging their methods of changing nucleus size?

This world of understanding will be rife with old dangers. In some sectors, this sort of knowledge, this ever-increasing number of diagnostic labels, will probably guarantee more reasons why the poor or poorly connected will be denied jobs or fair housing or health insurance. Even more menacingly, with scientific understanding comes the potential for manipulation, and the temptation to judge and fix things that aren't broken is never far behind. In every generation, there will be brownshirts whose biological ideal comes with an Aryan profile, and in every generation there will be some scientists and physicians who will be happy to march into hell with them.

But this new knowledge will be rife with promise as well. Recognizing the continuity between the workings of our benign little personality quirks and the versions that might qualify as disease would benefit those with the latter. When science teaches us repeatedly that there but for the grace of God go I, when we learn to recognize kinship in neurochemistry, we will have to become compassionate and tolerant, whether looking at an illness, a quirk, or a mere difference. And when this recognition spreads, we will have learned that drawing a boundary between "the essence of a person" and "the biological distortion of that essence" is artificial. It is simply a convenient way to classify the biological limitations common to most of us and other, rarer limitations. Being healthy, it has been said, really consists of having the same disease as everyone else.

FURTHER READING

At the end of each piece, I will be including some annotated references. This is both to document claims or findings mentioned and to point readers in the right direction should they want to learn more about specific topics. Unfortunately, in most cases, the references cited are pretty technical. I will try to note the ones that are more accessible, and to warn about the ones that are anything but.

For an overview of schizotypalism, see M. Vollema and R. van den Bosch, "The Multidimensionality of Schizotypy," *Schizophrenia Bulletin* 21 (1995): 19. For discussions of how schizotypalism and schizophrenia are on a genetic continuum (i.e., often occur in the same families), see K. Kendler, A. Gruenberg, and D. Kinney, "Independent Diagnoses of Adoptees and Relatives as Defined by DSM-III in the Provincial and National Samples of the Danish Adoption Study of Schizophrenia," *Archives of General Psychiatry* 51 (1994): 456. Also see C. Webb and D. Levinson, "Schizotypal and Paranoid Personality Disorder in the Relatives of Patients with Schizophrenia and Affective Disorders: A Review," *Schizophrenia Research* 11 (1993): 81.

For a review of the behavioral features of Huntington's disease, see J. Cummings, "Behavioral and Psychiatric Symptoms Associated with Huntington's Disease," *Advances in Neurology* 65 (1995): 179. Also see M. Mendez, "Huntington's Disease: Update and Review of Neuropsychiatric Aspects," *International Journal of Psychiatry in Medicine* 24 (1994): 189.

Frontal disinhibition syndromes are reviewed in J. Cummings, "Frontal-Subcortical Circuits and Human Behavior," *Archives of Neurology* 50 (1993): 873.

Tourette's syndrome is covered in O. Sacks, *The Man Who Mistook His Wife for a Hat and Other Clinical Tales* (New York: Summit Books, 1985). This is a highly readable collection of case studies of patients of Sacks's, an acclaimed neurologist.

Temporal lobe personality features are covered in S. Waxman and N. Geschwind, "Hypergraphia in Temporal Lobe Epilepsy," *Neurology* 24 (1974): 629. The late Geschwind, one of the giants of neurology, formalized much of the thinking about temporal lobe personality.

A wonderful introduction to obsessive-compulsive disorder can be found in J. Rapoport, *The Boy Who Couldn't Stop Washing* (New York: Signet, 1989). The author, probably the leading authority on the subject, has assembled some fascinating case histories.

Simon Levay's work on the biological bases of sexual orientation is covered in a number of publications. His key paper concerning structural differences in the brain is S. Levay, "A Difference in Hypothalamic Structure between Heterosexual and Homosexual Men," *Science* 253 (1991): 1034. He reviews the subject in a popular book, S. Levay, *The Sexual Brain* (Cambridge, Mass.: MIT Press, 1993). Levay's findings and, to an even greater extent, work by Dean Hamer concerning the possible genetic bases of sexual orientation remain controversial. The features of that controversy are summarized in a very accessible debate: S. Levay and D. Hamer, "Evidence for a Biological Influence in Male Homosexuality," *Scientific American* 270 (May 1994): 44; and W. Byne, "The Biological Evidence Challenged," *Scientific American* 270 (May 1994): 50.

Primate Peekaboo

Baboon Doc Sez: "Everyone Likes to Watch"

*Beryl Cook, Keyhole, 1981;
courtesy Portal Gallery, London*

Already, it's all beginning to blur, although I'm trying to hold on to the details. Was "Kato" Kaelin secretly Tonya Harding's lover? Do I remember correctly that Heidi Fleiss hired the Menendez brothers to do the hit on John Wayne Bobbitt? And was Woody Allen driving the white Ford Bronco when Michael Jackson fled the apartment being rented by John-John Kennedy and Daryl Hannah?

It's embarrassing to have such a poor memory sometimes. And it's more embarrassing to have ever known such things. You sit there on the commuter train, conspicuously reading Sartre in French, Sony Walkman cranked up so everyone can hear that you're listening to a Bartók string quartet. And then you can't help it, you just can't stop yourself, you have to lean forward ever so slightly to peer over the shoulder of the person in front of you, the one with the *People* magazine, all to check out Leslie Abramson's new hairdo.

It's even worse when you have a get-together of friends. There's a tentativeness about bringing up any of the hot scandals, an offhandedness. Once the subject is raised, there are definitely rules on how you can do it, as a card-carrying intellectual. Simpson—America, where you can get all the justice that money can buy, versus the racial angle, versus another go at the death penalty debate. Woody and Mia—breakdown of the nuclear family. The night wears on and the pressure to be detached and intellectual grows. Lorena Bobbitt—obvious parallels to Australian aboriginal genital mutilation rites. Menendez brothers—new unlikely insights into the Louisiana Purchase and the decline of the American bison herds. And despite the highbrow tone, you still feel sullied. *We should be discussing Marx and what his carbuncles had to do with dialectical materialism.* And because of the highbrow tone, you feel maddeningly frustrated. *I want to ask everyone who they think would be more of a drag to be stuck with in an elevator, Johnnie Cochran or Nancy Kerrigan.*

How can we be so voyeuristic, so little able to resist sensationalism and Oprahatics? When I've hit bottom, when I'm exhausted at work because I stayed up late watching Ted Koppel—even Ted!—go after one of these fifteen-minutes-of-famers, I resort to a reassuring fact. We are not alone.

I lead a dual scientific life, and when I'm not in the laboratory studying neurons and stress hormones, I spend my time in the grasslands of East Africa, studying troops of wild baboons. I spend an inordinate amount of time watching their social machinations, the complexities of their interactions among a hundred animals in the savannah. And they do the same thing.

Bunch of baboons sitting around in a field when there's a fight. Two large, high-ranking males, tension has been building up between them over something, and it finally erupts. A hundred pounds of muscle and testosterone, sharpened canines that are bigger than in an adult lion, slashing, lunging, brawling. Someone in the vicinity might get hurt, either amid the fighting or immediately afterward, with the loser taking out his frustrations on someone smaller. What's the logical thing? Get the hell out of there. And what do half the animals do? Stop what they're doing, stand bipedally, push in closer, all for a better view. Perhaps they are trying to pick up some pointers about strategy, or maybe they are bearing silent witness to the failure of pacifism. Nah. They just want to see what's going to happen.

Sometimes the voyeurism takes another familiar form. A few years ago, an adolescent male whom I named Absolom joined the troop, and he took this habit to new heights. He had just discovered girls, i.e., female baboons. Nothing was happening to him personally in that department, and he was reduced to a vigilance that consumed half his day. *Any* sexual consortship in the troop, and he would be lurking around in the bushes nearby, trying to catch sight of the good stuff, craning for a view of the action, holding his tail throughout. Here's how bad he got: one day, a high-ranking male and a female who was at the peak of being in heat sat quietly grooming each other. They were peripheral to the rest of the troop, secluded, no doubt working up to a moment of even greater intimacy. Suddenly, Absolom, who had silently slithered his way out to a branch of the tree just above them for a really good view, collapsed it under his weight and crashed down on top of them. None were pleased.

And sometimes the voyeurism has the feel of small-town coffee klatching. It was the season that a female named Rebekah had her first child. Primiparous mothers—those with their first child—are rarely particularly skillful, but Rebekah was plain awful. She forgot the kid when she left a group of other females, slapped him frequently, couldn't seem to learn how to position him to ride on her back, so that he sprawled sideways, clutching the base of her tail. One day, as she leapt from one branch to another in a tree with the kid in that precarious position, he lost his grip and dropped ten feet to the ground. We various primates observing proved our close kinship, proved how we probably utilized the exact same number of synapses in our brains in watching and responding to this event by doing the exact same thing in unison. Five female baboons in the tree and we two humans present all gasped as one. And then fell silent, eyes trained on the kid. A moment passed, he righted himself, looked up in the tree at his mother, and then scampered off after some nearby friends. And as a chorus, we all started clucking to each other in relief.

An anthropologist once said, "Humans had to invent language so we would have something to talk about around the fire at night." Koko and Michael are the gorillas famed for having been taught the rudiments of American Sign Language. Whether that constitutes language use on their part is rejected by most in the know, and I find their criticisms pretty convincing. Nevertheless, they are doing something communicative. Once, one morning, Michael witnessed one of his human teachers arguing with his girlfriend. And later that day, Michael told another human teacher about it. Proto-gossip.

Mark Twain defined humans as the only species that can blush or that needs to. We may indeed blush when we reveal that we're unduly informed about the details of the Presley-Jacksons, whereas the baboons would never blush as they elbow for a better view of their equivalent of Madonna tussling with David Letterman. But beyond that, there are few differences. Rubber necks appear to be a common feature of the primate order.

EPILOGUE (SUMMER 1996)

There was a temptation to update this piece by substituting this week's scandals. The trouble is, of course, that this week's and last year's will be equally archival by the time next week rolls around, so, by definition, it is not possible to have this piece be anything other than a trivia test by the time it is published. Thus, to refresh the memories of most readers, or to give important new information to those who wasted their time a few years back thinking about global warming or the collapse of the Soviet Union:

Kato was the world's most famous houseguest and poster child for the Let's Put Diction Lessons Back in our Classrooms movement. Tonya and Nancy provided us with the greatest villain/heroine set piece since Saturday cliff-hanger movies until Tonya pushed the villainess envelope by crying in front of the Olympic judges and Nancy did in her statuesque goddess number by being mean and cranky at Disneyland. Meanwhile, back on the farm, Heidi Fleiss illegally provided a service so that a bunch of rich and famous Hollywood males with testosterone problems could do something illegal in return and wound up being the only one to get in trouble for it.

Leslie Abramson inspired future generations of lawyers by sticking the Menendez boys in cardigans and temporarily delaying their getting convicted for blowing away their folks; she at least resisted the "take pity on them, they're orphans" gambit. In an unrelated episode, Lorena Bobbitt took an ax and gave John Wayne forty whacks, or at least one effective one; judging by his subsequent behavior, it's clear he didn't learn much from the incident.

Woody and Mia gravely disappointed us all by winding up on this list.

Michael Jackson possibly did some disturbed things with children and then, just when we least expected it, managed to top that. The Presley-Jacksons' connubial bliss lasted about a year, if my history textbooks are correct. Meanwhile, JFK Jr. and Daryl Hannah had such a brief, incandescent instant of being The Most Important Couple in the World before giving the crown back to those fun Windsors that I can't remember a thing about them anymore. Sorry.

On another front, the white Ford Bronco was part of the pioneering marketing gimmick of generating sales by having celebrities drive it slowly. And Johnnie Cochran, besides fostering racial tolerance in our land, pioneered the use of wool caps and rhyming couplets in the courtroom. Finally, Madonna appeared on David Letterman's show and was foul and unappealing even by late-night TV standards. The incident did not particularly harm her: as of this writing, she is reveling in her next supernova of fifteen minutedom with her first foray into maternal behavior, while continuing to serve ably as an associate justice of the U.S. Supreme Court.

FURTHER READING

People magazine, of course.

The incident about Michael the gorilla is reported in T. Crail, *Apetalk and Whalespeak* (Los Angeles: J. P. Tarcher, 1981), 137, 150.

The Night You Ruined
Your Pajamas

Scherer and Ouporov, Jacob's Ladder, 1994;
courtesy the artists

Oh, what an unmitigated horror puberty was. For girls, there was the inevitable first menstruation that came right in the middle of gym class: the bloody shorts, the panic—no one had told you you'd bleed that much—and the conviction that you were about to die of either embarrassment or hemorrhage. Training bras, pubic hair, acne; there was no end to disquieting changes. For boys, there was the first wet dream, inevitably occurring at sleep-away camp—sneaking off to the woods to bury your disgusting soiled pajamas—the visible erection with a mind of its own in the swimming trunks, the voice that cracked whenever you spoke to someone of the opposite sex. Humiliation, agonies, pimply insecurities; you wondered if life would ever return to normal, and the answer, you knew, was no. Most of all, there was the heartfelt prayer of every pubescent child: "Please, please, if I have to go through this, don't let it happen to me one minute before or after it happens to everyone else. Please don't let me be different."

For the peer-pressured adolescent, the timing of puberty is everything. It turns out that for all mammals going about the business of being fit enough to survive and reproduce, the timing of puberty is pretty important. And some mammals do something about it. As social or environmental conditions shift—when there's a change in the availability of food, for example, or in the number of animals competing for mates—many animals find that it sometimes pays to reach puberty at an earlier age, and it's sometimes better to delay the process.

Now there's evidence that humans, too, might alter the timing of puberty's onset. In a controversial paper published in 1991, a team of researchers proposed that girls reach puberty at an earlier age if they are raised in homes filled with parental strife or where the father is absent because of divorce or abandonment.

The notion that our bodies can engage in such weighty deliberation shouldn't be surprising. After all, we're comfortable with the idea that some of our lives' landmark events are decided strategically. We do it constantly. For example: "Oh, let's not. What a terrible time this would be for me to get pregnant—you're out of work, my dissertation research is bogged down, the Nazis are marching on Paris and we'll have to flee." That sort of thing.

Such strategies, however, are conscious. It's harder to think about the timing of puberty as a conscious strategy. We do not think, "The time is ripe—I've decided to start ovulating," any more than a bamboo plant decides, "The rainfall's been fabulous—I think I'll start flowering," or a deer decides, "Ah, spring is in the air—time to grow me some antlers." The "thinking" and "deciding" do not really occur. Instead, a more complex evolutionary process is taking place: organisms that have the biological means to time an important life-history event (such as reaching puberty or flowering or growing antlers) so that it occurs at an optimal time have an advantage over organisms that cannot do so; therefore more of them will survive, more will reproduce, and more will leave copies of their genes to future generations. And thus their adaptive timing mechanism will become more common in their species over the millennia.

When does it make the most sense to go through puberty? Well, to start with, puberty is an energetically costly experience, so it is logical to go through it early only if you've been fed well and you're healthy and can afford the metabolic expense. It wouldn't hurt to make sure that there is someone attractive around who is available to mate with, and it also helps if the environment is conducive to the survival of any children you may have. And when is it logical to defer puberty? If things are bad enough that you need your energy to survive instead of to reproduce, if it's logical to still be taken care of by Mom instead of becoming a mom yourself, if there's no one around to mate with except close relatives.

Consider, as an example, a female prepubescent mouse left on her own. Eventually, of course, she will reach puberty. But put an adult male into the cage with her and she will reach puberty earlier. This intriguing phenomenon, known as the Vandenbergh effect, is produced by pheromones, odors in the male's urine that cause a behavioral or physiological response in other animals. These pheromones, when smelled by the female, crank up her ovulatory machinery. The male's urine does this trick even when the male isn't present—just dab a smidgen of male urine on females and, voilà, earlier puberty. Moreover, the more male sexual hormones the mouse has in his bloodstream, the more effective he is at accelerating the onset of puberty.

Conversely, when a prepubescent female mouse is exposed to a large number of adult females, puberty is delayed—why bother if there are no guys around? Once again the signal is a pheromone, found in the urine of the adult females. Block the young female's sense of smell, and the puberty delay no longer occurs.

Researchers have speculated that these signals might play some role in regulating population density. For example, a large number of adult females probably indicates a dense population. The puberty-delaying pheromones serve as a brake on fecundity: if the population density is high enough, there will soon be food shortages, and why should a starving female waste energy on ovulating and getting pregnant when the odds of her carrying through a pregnancy are pretty small?

Such speculation is supported by a study conducted by zoologists Adrianne Massey at the North Carolina Biotechnology Center and John Vandenbergh (i.e., Dr. Vandenbergh Effect) at North Carolina State University. The two researchers examined mouse populations in the acre or so of grass found in each loop of a highway cloverleaf out in the hills of North Carolina. Patches like these constitute "biogeographic islands," closed mini-ecosystems where there are few mice immigrants (the ones who try to switch to a different cloverleaf patch usually wind up as roadkills). Massey and Vandenbergh studied the fluctuation of mouse populations in each patch. When the population of a particular patch became large enough, the females started to produce the puberty-delaying pheromone. And when the population dipped, the females stopped making this pheromone. Males indiscriminately made the puberty-accelerating odorant all the time, so when population density dropped and the delaying signals from the females were absent, puberty was accelerated. The bodies of prepubescent females were able to monitor the environment around them and choose wisely as to when to leap into the reproductive business.

Male mammals often show a similar savvy. Antelopes such as gazelles and impalas grow up in social groups in which a number of females and their offspring live with a single breeding male. Other mature males live either as wandering loners or in all-male bands, butting heads and honing their fighting skills for the moment when they stage a coup d'etat against the breeding male.

For a male growing up with his mother and the rest of the

breeding group, puberty carries a stiff price. When an adolescent male sprouts pubescent "gender badges" such as horns, the breeding male perceives the youngster as a potential sexual rival and harasses him, driving him out of the group. The apron strings are cut rather abruptly, with consequences that are not trivial—when males go out on their own, their risk of being eaten by predators soars.

What is a prepubescent male antelope to do? By reaching puberty too early, he is subject to the abuses of the dominant male when he may not yet be ready to survive the rigors of exile from hearth and home. But by delaying puberty too long, he forgoes reproductive potential. Clearly he needs to make a carefully considered decision—and apparently he does. Richard Estes of Harvard's Museum of Comparative Zoology has examined the timing of puberty in antelopes and found that if the outside world looks pretty challenging and dangerous, with other antelopes around to fight over limited food and a lot of lurking predators, the males delay the onset of puberty.

Similar strategizing occurs in primates. If you have ever watched orangutans in a zoo, you may have noted that the males seem to come in two varieties. One is short-haired, limber, and agile. The other is much heavier and lumbering, with these weird fatty cheek flanges, thick long hair, and a huge muscular throat pouch. Primatologists studying orangutans in the wild always assumed that the more gracile form was simply an adolescent, the latter being the adult version. However, in zoo populations, some males retain the gracile form for decades. Invariably these are males living near an older, more socially dominant male. Some sort of signal given off by the dominant male makes the body of the subordinate delay the development of the adult secondary sexual features that give the heavier form its appearance. Take away the dominant male, and the eternally youthful subordinate rapidly develops all the mature traits.

None of this strategizing has yet been shown to occur in male humans, but psychologist Jay Belsky and anthropologist Patricia Draper of Penn State and psychologist Laurence Steinberg of Temple University believe they have found something very much like it in human females. Their argument is actually the flip side of the delaying tactic used by antelopes and orangutans. In a nutshell, their premise is: When times are tough and life is unstable, hurry up, reach puberty, and start reproducing sooner.

This thinking bears the mark of a certain style of ecology. One of the basic dichotomies made by ecologists is between "stable" and "opportunistic" species. Stable species live in unchanging, predictable environments. They live long lives in large populations, they are large-bodied, and they lavish a lot of parental attention on relatively few offspring. They opt for quality over quantity when it comes to reproduction; years later, members of these species are still doing the kids' laundry and paying tuition for chirping lessons. Compared with closely related species that don't have as many of these traits, stable species tend to reach puberty later. For them there is no rush: they can afford to grow big and healthy before starting to invest all that energy in a few pregnancies. Think of elephants, with their two-year pregnancies and sixty-five-year life spans. Or ancient redwood trees. Or humans, a classic stable species.

In contrast are the opportunistic species, which live in ecosystems that are unstable and unpredictable, with long periods of severe weather, food shortages, and environments that are favorable only now and then. Animals that live near bodies of water that dry up seasonally or plants that grow best in soil just after a fire are opportunistic. They're gamblers, living boom-or-bust existences. The population will be tiny for long periods. Then, when conditions suddenly become right, there's tremendous pressure to take advantage of them fast, and, wham—everyone starts mating or pollinating. Quantity over quality, reproduce as much as possible, don't bother taking care of the kids, just flood the market with offspring while the going is good. Opportunistic species tend to be small-bodied and short-lived, their population size fluctuates wildly, and once conditions are favorable, they reach puberty rapidly. Think of rabbits reproducing like rabbits and invading some new region, or of some kind of scrub weed that pops up overnight in the disturbed soil on the edge of a construction site.

About a decade ago, Draper and her colleague Henry Harpending theorized that when girls are abandoned by their fathers during childhood, they grow up to become women who behave more like an opportunistic species. The girls have learned that males make unreliable parents, so they don't bother searching for a reliable mate—they have sex at an earlier age, have more sexual partners, more kids, and are more likely to have rocky marriages.

When Draper and Harpending proposed their theory, it was already well known that women who grow up in unstable homes are indeed more likely to behave this way—the novel thing about Draper and Harpending's work was that they tried to explain this pattern from an ecological and evolutionary perspective rather than a psychological or socioeconomic one. Now Draper, Belsky, and Steinberg have built on this hypothesis by making the controversial suggestion that in unstable settings it's not only the *behavior* of such girls that becomes more opportunistic, but their *bodies* as well, and so the girls reach puberty earlier.

A handful of studies supports this idea. In two studies led by Belsky and one by Steinberg, girls who reported more strained relations with their parents, and girls raised in homes with persistent family conflict, reached puberty earlier than girls from more contented families. In a detailed 1990 study, Michele Surbey of Mount Allison University in Canada (who at the time proposed a theory quite close to that of Belsky and colleagues, but got no attention in the media for it) found that girls raised with the father absent because of divorce or abandonment also had earlier puberty. An Australian research group had found the same thing in 1972.

The various authors wrestle a bit with potential physiological mechanisms for this effect (evolutionary theory is concerned with *why* puberty might be accelerated; physiological mechanisms, the nuts and bolts of hormones and the nervous system, explain *how* puberty might be accelerated). Surbey hypothesizes that biological fathers might release a puberty-delaying pheromone; take the father and his pheromones away and you've got earlier puberty. The problem here is how to explain Belsky's results, in which the presumably pheromonally smelly father sticks around and constantly fights with the mother, with the same result of earlier puberty in the kid.

Regardless of what the mechanism might be, the mere idea that social instability leads to earlier puberty has upset a lot of people. Some investigators who have tried to duplicate these studies haven't gotten the same results. Other critics note that measures of "parental strife" or "strain" in girls' relations with parents rely largely on the girls' own reports, so there's plenty of room for bias and inconsistency.

The biggest problem with the findings and the underlying theory is that they are exactly the opposite of much of what is known about the effect of stress on puberty. A lot of the research suggests that "when times are tough and life is unstable, hold off on reaching puberty—it's a bad time to get pregnant and the kid won't survive, so don't waste energy ovulating." And we're not just talking about antelopes here. Girls who experience the stress of sustained physical exercise (for example, serious ballet dancers) reach puberty later than average, as do girls with anorexia nervosa. The workings of this effect are pretty well understood—your body has to have a certain amount of fat deposited before it does something as risky and expensive as starting to ovulate on a regular basis, and girls whose muscle/fat ratio falls below a certain critical set point delay the onset of puberty.

Additional evidence that extreme physical stress suppresses reproductive physiology instead of stimulating it comes from studies of women who have already reached puberty. Consistently, physical stresses such as weight loss, heavy exercise, and illness delay or even block ovulation—again, muscle/fat ratios probably play an important role here. Moreover, psychological stresses can also inhibit ovulation. This is particularly relevant, given that the girls growing up in unstable home environments were probably being psychologically, but not physically, stressed.

So most investigators find that unstable environments filled with physical or psychological stressors inhibit reproductive physiology. Yet these new reports suggest that instability accelerates puberty onset in girls.

The authors of the new theory try to reconcile these differences. If Surbey is right that the girls reach puberty earlier because some puberty-delaying pheromone from their fathers is missing, then this is no longer a contradictory story about stress and maturation but instead a story about pheromones and maturation. Her data also suggest the broader theory might not be about "when life is unreliable, reach puberty earlier," but merely about "when the primary male around you is unreliable, reach puberty earlier." Surbey found that having a mother rather than a father absent didn't accelerate puberty, yet growing up without a mother should be just as much of a lesson to a child that life is stressful.

Belsky contends that there could be a difference between the effects of extreme, potentially life-threatening stress and milder forms of stress, such that a starving anorexic might delay puberty, while a stressed-out yet relatively healthy child of a broken home might reach it sooner. Yet experimental work shows that even milder stressors inhibit reproductive physiology (although to a lesser extent) rather than stimulate it. Another of Belsky's speculations would sidestep a lot of the perceived problems with his findings—maybe the girls in the unstable environments become depressed, eat more, deposit more fat, and thus hit puberty earlier, tying into the muscle/fat ratio feature of puberty onset cited above.

One other important factor may explain the effect, although it's messy and complicated. The age of puberty onset is partially inherited—mothers who reach puberty early tend to have daughters who do so as well. Perhaps early sexual maturation is somehow related to having unstable marital relationships later (since early sexual activity is). In that scenario, girls in unstable homes reach puberty earlier because they inherit that biological trait from their mother, not because the home environment is unstable. That would be a very different story, taking away a lot of the novelty of these findings. Surbey did find that the mothers of her early-maturing girls had indeed reached puberty early themselves. When the mother's age of puberty onset was taken into account, the overall effect of home environment decreased.

The ideas of Belsky, Surbey, and their colleagues may or may not turn out to be valid—now's the time to trot out the cliché about more research being needed. In the meantime, it's a perfect subject for a scientific disagreement, since there's a lot of minutiae to squabble over. But there's one problem that should have taken some of the pleasure out of the squabbling—the effect is tiny. When the various studies supporting this effect are averaged, they suggest that instability accelerates the onset of puberty in these girls by about five months. That's nothing; these kids can't even break in a new pair of jeans in that time. So why is it that not only are scientists bickering over this, but the lay press has picked up on this story?

Lurking behind the scientific arguments is a big public policy conflict over what to do about teen pregnancies. The pregnancy rate for teens in the United States is astonishingly high for a Western country, twice that of Canada or England, for example. Teen mothers come disproportionately from poor inner-city populations rife with unstable, broken families. Children of such mothers are more likely to have health and learning problems, and thus to underachieve and place a burden on society for years to come.

Belsky thinks the theory he's helped formulate has something to say about this mess. In his view, the unstable early environment of many an inner-city teenage girl produces earlier puberty and the whole range of attendant opportunistic behavior that ends in an increased risk of pregnancy. "These teenagers aren't necessarily carrying out bad behavior," he says. "Their bodies are just responding to forces that have emerged over the course of evolution."

Belsky and Surbey's critics, however, worry that the theory suggests that one can blame biology instead of society for the problem, and to some, blaming biology implies that we can't do much about teen pregnancy. For a bloatedly wealthy country, they note, we expend very little effort to give a massive underclass any hope for jobs or education. Our mass media is permeated with a sexual hard sell that few adolescents have the means to respond to maturely. Our government wages war on sex education and access to contraceptives and makes it increasingly difficult for the young and the poor to have abortions. If these are the reasons why we have such an appallingly high rate of teen pregnancies among our poor, there is lots we can do about it. Or is it because early family conflict commands such girls to follow a biologically determined program of early pregnancy?

Belsky doesn't think the explanations are mutually exclusive. He believes that his theory is not about how biology can cause earlier pregnancy, but about how a biological drive can make the body of a young person from an unstable home more likely to respond to various societal pressures. This is what he calls a "nature-based theory of nurture." If true, it predicts that the effects on puberty onset of growing up in an unstable home would be stronger among poor kids than among middle-class kids. And if he is right—if early stress and resulting biological changes do not form a direct link to more teen pregnancies but instead are one of the many factors influencing how teenagers respond to the seductive world around them—then perhaps the policy implications of his view are not so worrisome. Much of behavioral biology works through this sort of biology/environment interaction. These results are not an excuse for society to say, in effect, "There goes that darn biology again—nothing we can do about it," and wash its hands of the mess. The findings are, if anything, a scientific rationale for why society has the responsibility to try to solve these social ills.

EPILOGUE

Surbey's data were published in 1990; Belsky and colleagues published their more visible version of the same in 1991. What has happened since then? Have researchers been able to replicate the findings in their own study populations? Have the ideas contained in these papers influenced the thinking of other researchers?

To date, I am aware of only a single study that has replicated the finding that growing up in an unstable home environment leads to slightly earlier puberty onset in girls. While this is nice, Belsky was one of the coauthors, and one hopes to see what is called "independent" replication of a finding. This is where someone completely unconnected to the discoverer of a new fact, someone with no stake in the matter—even better, someone who is skeptical about this new fact—comes forward and says, in effect, "I thought there was no way that things worked like this, but we tried it, and you know what, they're right." That hasn't happened yet. That particular paper debunked Belsky's speculative chain where unstable home environment leads to depressed kids leads to more eating leads to more fat deposition . . . and earlier puberty (and that chain was unlikely to work, as an unstable home setting doesn't necessarily lead to depression, and depression is at least as likely to lead to undereating as to excessive eating). Note that that was Belsky as one of the coauthors reaching the conclusion that his earlier speculation wasn't right—that's a sign of an honorable scientist at work. The paper, instead, garnered support for Surbey's notion that the early puberty/unstable home life link was an artifact, and that the real link was women with early puberty onset having daughters with earlier onset. I suspect that that is the explanation, which would reconcile this small effect with the extensive earlier literature showing the antireproductive effects of a stressful environment.

What has been the influence of these studies on the thinking of others in the field? An interesting pattern has emerged. These studies have had some impact on the thinking of psychologists and evolutionary biologists about growth and development; the work has been cited about fifteen times by such scientists in recent years. In contrast, the papers have been entirely ignored by endocrinologists—in other words, scientists are often loath to pay attention to the work of individuals outside their immediate domain, even if it concerns work within their domain.

And naturally, as we speak, teen pregnancies continue unabated in the United States, untouched by this particular academic hiccup.

Final note: I kind of liked the title of this piece until someone canvassed a bunch of friends and discovered that they thought the title was male-biased. I was thoroughly puzzled by this. While males' first nocturnal emissions are certainly likely to occur nocturnally, it seems obvious that the onset of one's menses can occur while in pj's as well (in humans, for example, there is a strong bias toward women giving birth at night, a pattern seen among most primates, who would just as well not give birth midday in the middle of a savannah filled with hungry predators). Time for some quantitative data. Does anyone know of any study concerning time of day of menses onset? Operators are standing by for your call.

FURTHER READING

The original reports can be found in M. Surbey, "Family Composition, Stress, and the Timing of Human Menarche," in T. Ziegler and F. Bercovitch, eds., *Socioendocrinology of Primate Reproduction* (New York: Wiley-Liss, 1990), 11; and in J. Belsky, L. Steinberg, and P. Draper, "Childhood Experience, Interpersonal Development, and Reproductive Strategy: An Evolutionary Theory of Socialization," *Child Development* 62 (1991): 647.

The follow-up study noted in the epilogue is T. Moffitt, A. Caspi, J. Belsky, and P. Silva, "Childhood Experience and the Onset of Menarche: A Test of a Sociobiological Model," *Child Development* 63 (1992): 47.

An introduction to stable and unstable species (known for agonizingly obscure reasons having to do with mathematical formulas as "K" and "r" species—the reasons even including the stricture of the "K" being capitalized and the "r" not) can be found in the classic E. O. Wilson, *Sociobiology: The New Synthesis* (Cambridge, Mass.: Harvard University Press, 1975). This is also the best book for an introduction to thinking about behavior having evolved strategically (i.e., the sort of issues discussed early on in the piece, such as when does an animal "decide" that it's a good time to carry out some behavior).

Work concerning unstable home environments and earlier onset of sexual behavior is summarized in P. Draper and H. Harpending, "Father Absence and Reproductive Strategy: An Evolutionary Perspective," *Journal of Anthropological Research* 38 (1982): 255.

A review of the Vandenbergh effect can be found in K. Darney, J. Goldman, and J. Vandenbergh, "Neuroendocrine Responses to Social Regulation of Puberty in the Female House Mouse," *Neuroendocrinology* 55 (1992): 534. The particular cloverleaf study is described in A. Massey and J. Vandenbergh, "Puberty Delay by a Urinary Cue from Female House Mice in Feral Populations," *Science* 209 (1980): 821.

A review of how stress generally inhibits reproductive physiology and behavior can be found in chapter 7 of R. Sapolsky, *Why Zebras Don't Get Ulcers: A Guide to Stress, Stress-Related Diseases, and Coping* (New York: W. H. Freeman, 1994). For a review of what fat deposition has to do with puberty onset, see R. Frisch, "Body Weight, Body Fat and Ovulation," *Trends in Endocrinology and Metabolism* 2 (1991): 191.

Measures of Life

❦

Francisco José de Goya, The Third of May, 1808, *1814;
Prado, Madrid, Alinari/Art Resource, New York*

I t is one of the most riveting paintings ever made. In 1808 the populace of Madrid rose up against an occupying French army, an uprising that was soon crushed; the painting depicts the mass executions of insurgent Spaniards that began before dawn on the third day of May. Goya captures the appalling, machinelike process of the slaughter, implied by the piled-up bodies and the waiting line—those men in the light, seconds away from death, who have time only for a single burst of emotion: mute terror, fervent prayer to an indifferent god, or a final bellow affirming life and existence. And then they too are dead. It is impossible not to think of their dying and impossible not to imagine oneself in that circumstance. One wonders: How would I react? Would I be brave? Again and again one's eyes return to the men in that circle of final light.

It is only later that the mind returns to the other men in the picture. The faceless ones, automatons with guns, the members of the firing squad. Although their pull on the viewer is more gradual than that of the doomed men in the light, it is just as powerful. For those French soldiers were the agents of one of the most magnetic of human dramas, the taking of life. How could they do it? What did they feel? Did they feel? Perhaps they were evil, or perhaps they were coerced. Or perhaps they truly believed in the necessity of their actions.

One of the great horrors of a world that provides its Goyas with endless such scenes to paint is the possibility that after the smoke clears and the corpses are removed, the executioners give the executions no thought whatever. More likely, though, at least some of those faceless men do reflect on their work, and do feel guilt over it or at least fear that others will someday judge them as well. In fact, such remorse seems pervasive enough that firing squads have evolved in a way that, in some circumstances, accommodates it. Central to that evolution are some subtly distorting aspects of human cognition that have much to do with how people go about killing people, how they judge people, and how they set priorities for their resources. And oddly, such cognitive distortions may have something to do with how some of us feel about doing science.

Underpinning these suggestions is work done over the past twenty years, principally by two psychologists, the late Amos N. Tversky of Stanford University and Daniel Kahneman of Princeton University. Tversky and Kahneman have shown that one can present people with two choices that, in terms of formal logic, are equivalent, yet one choice may be strongly and consistently preferred and may carry an emotional weight altogether different from that of the other choice. On the other hand, one can offer people two logically disparate choices, yet they may be seen as equivalent. Such apparent contradictions could come about in several ways: they could be caused by some commonly held cognitive biases identified by Tversky and Kahneman; they could be the outcome of the way the choices are presented; or they could be a reflection of the makeup of each individual.

Consider this scenario: You are told about a young woman who in college was much involved in leftist and progressive causes, a true social activist, committed and concerned. You are then told that the young woman has since gone on to one of four careers, and you are asked to rate the likelihood of her having pursued each one: (1) an organizer of farmworkers (good chance, you say); (2) a bank teller (seems highly unlikely); (3) an environmental activist (another likely outcome); and (4) a bank teller who is active in the feminist movement (well, surely that is much more likely than her having become a mere bank teller; at least she's politically involved some of the time). Thus you rate choice 4 as being more likely than choice 2. Logically, however, it is impossible for choice 4, which features two constraints (bank teller and activist), to be more likely than choice 2, which has only one of those two constraints.

Much of Tversky and Kahneman's work has gone into figuring out why people sometimes make that distortion, embedding something more likely inside something less likely. Another aspect of their work identifies ways people regard scenarios as unequal when in fact they are formally equivalent:

> *You are a physician and you have a hundred sick people on your hands. If you perform treatment A, twenty people will die. If you opt for treatment B, everyone has a 20 percent chance of dying. Which one do you choose?*

The alternative scenario runs like this:

> *You have a hundred sick people. If you perform treatment A, eighty people will live. If you opt for treatment B, everyone has an 80 percent chance of living. Which one do you choose?*

The two scenarios are formally identical; they are merely framed differently. But it turns out that for the first scenario, which states things in terms of death, people will prefer option B, whereas for the second scenario, stated in terms of survivorship, people prefer option A. Thus when thinking about life, people prefer certainty; when thinking about death, they prefer odds, because it is always conceivable the odds can be beaten.

What if Tversky and Kahneman's scheme were applied to an early-nineteenth-century militia bent upon taking fatal revenge on some prisoners? At the time, a single shot fired from a moderate distance was often not enough to kill a person. Thus, some options: one man could stand at a distance and shoot at the prisoner five times; or five men could stand at the same distance and shoot at the prisoner once each.

Five-times-one or one-times-five are formally equivalent. So why did the firing squad evolve? I suspect it had something to do with a logical distortion it allowed each participant: if it takes one shot from each of five people to kill a man, each participant has killed one-fifth of a man. And on some irrational level, it is far easier to decide then that you have not really killed someone or, if you possess extraordinary powers of denial, that you have not even contributed to killing someone.

Why do I think the firing squad was an accommodation to guilt, to the perception of guilt, and to guilty consciences? Because of an even more intriguing refinement in the art of killing people. By the middle of the nineteenth century, when a firing squad assembled, it was often the case that one man would randomly be given a blank bullet. Whether each member of the firing squad could tell if he had the blank or not—by the presence or absence of a recoil at the time of the shooting—was irrelevant. Each man could go home that night with the certainty that he could never be accused, for sure, of having played a role in the killing.

Of course, the firing squad wasn't always the chosen method of execution, and that is where Tversky and Kahneman's work says something about the emotional weight of particular killings. If civilians were being killed—members of an unruly urban populace rising up in threatening protest—a single close-range shot to the head, a bayoneting, or any such technique would be suitable. If the victim was some nondescript member of the enemy army, a firing squad was assembled in which there was rarely a random blank. But if, by the addled etiquette of nineteenth-century warfare, the victim was someone who mattered, an accomplished and brave officer of the enemy or a comrade turned traitor, the execution would be a ceremony filled with honorifics and ambivalence.

For example, military law for the Union army during the American Civil War specified the rules of executing a traitor or deserter. Explicit instructions were given to cover a range of details: how the enlisted men in attendance should be lined up to ensure that everyone would see the shooting; the order in which the execution party marched in; the music to be played at various points by the band of the prisoner's regiment—and exactly when the provost marshal was supposed to put a blank into one of the guns, out of sight of the firing squad.

It all seems so quaint. But the same traditions persist today. In the American states that allow executions, lethal injection is fast becoming the method of choice. In states more "backward" about the technology of execution, execution is done by hand. But among the cutting-edge states, a $30,000 lethal-injection machine is used. Its benefits, extolled by its inventor at the wardens' conventions he frequents, include dual sets of syringes and dual stations with switches for two people to throw at the same time. A computer with a binary-number generator randomizes which syringe is injected into the prisoner and which ends up in a collection vial—and then erases the decision. The state of New Jersey even stipulates the use of execution technology with multiple stations and a means of randomization. No one will ever know who really did it, not even the computer.

Such rites of execution are part of a subtle cognitive game. The formal set of circumstances exemplified by the multiple executioners who contribute to a singular execution has challenged societies in other forms. If five people must shoot to kill a man, then on some logical level no one of the shooters is a murderer. A different version of that principle is central to the idea of causation as taught in the law schools. Suppose two men start fires simultaneously, at opposite ends of a property. The fires merge and burn down the property. Who is responsible for the damage? The logic adopted by nineteenth-century American courts would have given solace to any participant in a firing squad. Each of the two arsonous defendants can correctly make the same point: If I had not set the fire, the property would still have burned. So how can I be guilty? Both of them would have walked free.

By 1927 a different interpretation rose out of a landmark court decision, *Kingston* v. *Chicago and NW Railroad.* The case did indeed begin with two fires—one caused by a locomotive and the other of unknown origin; the fires then joined to burn down the plaintiff's property. Before *Kingston,* judges would have ruled that the guilt of the singular burning of a property could not be distributed among multiple parties. But in the *Kingston* case, the courts declared for the first time that guilt for a singular burning, a singular injury, a singular killing, could be distributed among contributing parties. The decision prompted an almost abashed disclaimer from the judge: were the railroad able to get off free, he said, "the injustice of such a doctrine sufficiently impeaches the logic upon which it is founded." It was as if he had to apologize for being compassionate instead of logical.

His decision, however, was no more or less logical than the earlier ones that would have freed the railroad of wrongdoing. Again, it is a matter of Tverskian and Kahnemanesque framing: two fires join and burn down a property. The issue of guilt depends on whether the judges are more emotionally responsive to the defendants ("If I hadn't been there, the place would still have burned. How could I be guilty?"), or more responsive to the plaintiff ("My home was burned by these people"). Daniel J. H. Greenwood of the University of Utah College of

Law, who has thought long and hard about the 1927 philosophical transition, thinks the change reflects the general social progressiveness of the time: it shed some vestiges of the legal thought that had reflected the interests of robber barons and trusts. Judges began to frame cases logically in a way that would make them think first of the individual and damaged plaintiff, and that made it more difficult for anyone to hide behind corporate aggregateness. Suddenly it was possible for a number of people to be statistically guilty for a singular event.

The act of placing a blank among the bullets for the firing squad is even subtler in its cognitive implications. For example, on what occasions are people allowed the comfort of a metaphorical blank? Criminal law in the United States requires the unanimity of twelve jurors in a decision. You can bet that when the all-white jury acquitted the police in the first Rodney King trial, the jurors desperately wished for a system that would have allowed them to vote eleven-to-one for acquittal. Each one could then have hinted to the enraged world that he or she had been the innocent in that travesty. Yet the system does not allow such blanks, probably because the desire for perceived unanimity when the State forces its citizens to judge one another outweighs the desire of the State to protect the citizens who do the judging.

An even more intriguing aspect of the metaphorical blank is what one does with it cognitively. When a member of the two-man lethal-injection team goes home the night after the execution—and assuming he has a twinge of conscience— he does not think, "I am a killer," or "Today, I contributed to a killing," or even "I have a 50 percent chance of having helped kill someone today." More likely, he would frame the same logic in terms of statistical innocence: "I have a 50 percent chance of not having helped kill someone." Or he might even find a way of rationalizing a fraction of innocence into an integer: "One of us didn't really do it. Why shouldn't it be me?"

Or: "I know what the dummy button feels like. It was mine."
Given the duality of perceiving the event as one of fractional
guilt or fractional innocence, people not only bias toward the
latter, but also are replete with clever rationalizations that distort the matter further until it is the certain integer of innocence. In other circumstances the bias goes the other way. An
archetypally villainous industrialist, bloated and venal, con-
templates a new profitable venture for his factory. His min-
ions of advisers, on the basis of their calculations, tell him that
the toxins his factory will dump into the drinking water will
probably lead to three cancer deaths in his town of 100,000.
Naturally he decides to go ahead with his remunerative plans.
But that night he surely doesn't think in integers: "Today, for
profit, I have consigned three innocents to their deaths."
Instead, all his cognitive biases lead him to frame the matter
in terms of statistical guilt: "All it does is increase the cancer
risk .003 percent for each person. Trivial. Charles, you may
serve me my dinner."

Tversky uses the term "tendencies" to describe the practice
of conceiving guilt and innocence in fractions. He reserves the
term "frequencies" to describe the bias toward viewing the
same facts in integers. If one has done something bad, it is no
accident that one thinks fractionally—or, statistically speaking, in terms of distributed tendencies, a world without complete faces. If one has done something good, the pull is toward
frequencies and integers.

Thinking in frequencies is also the easiest way of getting
someone to consider another person's pains. It is overwhelmingly likely that if it is ever decided in the courts to hold
tobacco companies responsible for killing endless numbers of
people, the decision will not be a result of a class action suit,
with claims of distributed pain. Rather, it will be because a
jury will understand the pain of one individual who personi-
fies being killed by cigarettes. "People think less extensionally

when they think of tendencies than when they think of frequencies," says Tversky. In other words, as every journalist knows, empathy is grounded in a face, an individual story, a whole number of vulnerability.

The work of Tversky and Kahneman teaches scientists all kinds of lessons about cognitive pitfalls, including the fairly obvious ones people confront as they struggle with problems that invariably have multiple causes. For example, is the causality distributed among so many agents that people cannot perceive it? Do people distort their work with a cognitive bias toward finding a single magic bullet? We who do science all could learn a thing or two from Tversky and Kahneman. But there is an aspect of their work that tugs at my emotions. It comes from another of their scenarios:

> *In a population, all the deaths are attributable to two diseases, each accounting for half the deaths. You have a choice. You can discover something that cures all the known cases of one disease, or you can discover something that cures half the cases of each disease.*

By now it should be clear that the two options are formally equivalent: $1 \times 1/2$ is equal to $(1/2 \times 1/2) + (1/2 \times 1/2)$. Yet people show a strong bias toward curing all the causes of one disease. That kind of integer satisfies a sense of closure. In spite of our poetry and abstract theorems and dreams, our cognitive pull is toward being concrete and tangible—cross that disease off the list of things to worry about. And as a scientist, that is what plagues me emotionally.

In 1977 a group of biomedical scientists from the World Health Organization inoculated the populace of a town called Merca, in the hinterlands of Somalia, and thereby accomplished something extraordinary: they eradicated the last known case of smallpox on this planet. I think of that moment often, and always with the envy of knowing I will never accomplish something like that. I had a similar sense when I was a postdoctoral fellow at the Salk Institute for Biological Studies. I would occasionally see Jonas Salk at seminars and feel in awe of him, a scientist who knew an extraordinary closure.

I will never achieve that sense, not just because I am not as skillful a biomedical scientist, but also because of the way science is typically done these days. Now the scientific arena is one in which a coherent picture of a problem emerges from the work of teams of people in dozens of laboratories, in which diseases often have multiple causes, in which biological messengers have multiple effects; in which a long, meandering route of basic research might eventually lead to clinical trials. The chances that one person can single-handedly bring closure to a biological research problem grow ever more remote.

My own research is at an extreme of distributed causality. In my laboratory, I study how stress, and a class of hormones secreted during stress, can endanger neurons in the brain and make those cells less likely to survive neurological disasters such as cardiac arrest and seizure. In other words, I do not study so much how stress can damage the brain as much as how it may exacerbate the toxicity of neurological insults. It is by no means clear yet whether things really work that way, let alone how important a variable stress is. But if every scientific fantasy of mine comes true—if, inconceivably, every experiment I go near works perfectly—I will be able to demonstrate that stress is an exacerbating factor in modulating the neurological damage that affects hundreds of thousands of people each year. And if every crackpot therapeutic idea I have works, the new knowledge will lead to a way of decreasing the brain damage, at least a little bit, in each of those people.

Even in my fantasy world, though, stress would be only one of many statistical villains. It may be a factor in many diseases that cause brain damage, but, at best, it is only a small factor; its importance would be distributed over the huge number of cases. With deep knowledge of the problem, the best I could hope for is to bring about statistical good—in effect, to save a hundredth or a thousandth of a life here and there. That would be wonderful, indeed. Still, there is the pull of integers, the pull of whole numbers.

Few scientists will ever save or solve in integers rather than in fractions. Saving in integers is the realm of clinical medicine; there one deals with one person at a time. Moreover, saving lives in integers is the realm of an era of science that for the most part is past, when the lone investigator could conceivably vanquish a disease. Instead, in the present world we scientists have factors that modulate and synergize and influence and interact, but rarely cause. It is abundantly true that $1 \times 1/2$ is equal to $(1/2 \times 1/2) + (1/2 \times 1/2)$. But we have a cognitive bias toward the former, and we are in a profession that must exalt extreme versions of the latter.

That bias must plague us scientists when we try to justify why we do the work we do. I don't mean justifications we come up with in our grant proposals, the final paragraphs where we go on about the endless (and unlikely) benefits that will occur if the National Institutes of Health bless us with money for a few more years. I mean the justifications we come up with in the middle of the night, when we think about being in a profession that requires us to pour radiation down a sink or to kill animals, that calls on us to work so hard that the rest of our selves wither, so that we ignore our loved ones. All for that wisp of a hope of making a difference, to achieve that scientist's state of grace where your work will help vast numbers of people somewhere down the line.

It is that fantasy that fuels the way in which science must now be done: battling distributed villains, charging them with statistical guilt, laying the groundwork that might eventually lead to minute fractional victories. Sometimes in the middle of the night, you begin to wish that progress in your profession was of a different sort, something not so at odds with our fundamental cognitive bias. Sometimes, it would be nice to simply see the face of someone we helped.

FURTHER READING

Readers interested in Kahneman and Tversky's work might check the following:

D. Kahneman and A. Tversky, "Choices, Values and Frames," *American Psychologist* 39 (1984): 341. Also see: A. Tversky and D. Kahneman, "The Framing of Decisions and the Psychology of Choice," *Science* 211 (1981): 453.

In addition, a nontechnical summary of their work can be found in K. McKean, "Decisions, Decisions: Two Eminent Psychologists Disclose the Mental Pitfalls in Which Rational People Find Themselves When They Try to Arrive at Logical Conclusions," *Discover,* June 1985, 22.

Descriptions of modern execution technology can be found in S. Trombley, *The Execution Protocol* (New York: Crown Publishers, 1992).

The Young and the Reckless

Mark Tansey, Myth of Depth, *1984;*
Private Collection, New York, courtesy Curt Marcus Gallery, New York

I remember going off to college. I was so nervous and excited that I had the runs for a week beforehand. What if they had made a mistake by admitting me? Suppose I never made any friends. Would these really turn out to be the best years of my life? Queasy and vertiginous, I packed my bags onto a Greyhound bus, squirreling a bottle of Kaopectate into my knapsack.

Actually, getting my bowels into such an uproar proved to be quite justified, given the momentous freshman-year events that awaited me: the epiphany when I knew I was never going to understand photosynthesis and should give up on being a biology major; the realization, as I contemplated four years of purple yogurt and Polynesian meatless meatballs in the cafeteria, that my mother was a fabulous cook; my first lesson in political correctness—I learned that I was now surrounded by women and not by girls; the pleasant discovery that some of those women were willing to talk to me now and then; the wonder of watching older guys suavely work references to Claude Lévi-Strauss and Buckminster Fuller into casual conversation; the giddy pleasure at finding that a joke that worked in high school worked equally well here too; the calming ritual of fighting with my roommate every evening over whether to open or to close the window.

Growing up, and growing away. Off to college, off to war,

off to work in the city, off to settle in a new world—home is never the same again, and sometimes home is never even seen again. What is striking about this maturational event is that it is central not only to us humans but to many of our primate relatives as well. The process of growing up and growing away has a remarkably familiar look of excitement and discovery and challenge.

Some primates, such as orangutans, lead solitary lives, meeting only for the occasional mating. But the average group of primates is supremely social, whether it is a family of a dozen gorillas living in a mountain rain forest, a band of twenty langur monkeys on the outskirts of an Indian village, or a troop of one hundred baboons in the African grasslands. In such groups an infant is born into a world filled with relatives, friends, and adversaries, surrounded by intrigue, double-dealing, trysts, and heroics—the staples of great small-town gossip. Pretty heady stuff for the kid that's learning whom it can trust and what the rules are, along with its species' equivalent of table manners.

Most young primates are socialized in this way quite effectively, and home must seem homier all the time. Then, inevitably, a lot of the young must leave the group and set out on their own. It is a simple fact driven by genetics and evolution: if everyone stayed on, matured, and reproduced there, and if their kids stayed on, and their kids' kids too, then ultimately everyone would be pretty closely related. You would have the classic problems of inbreeding—lots of funny-looking kids with six fingers and two tails (as well as more serious genetic problems). Thus essentially all social primates have evolved mechanisms for adolescent emigration from one group to another. Not all adolescents have to leave. The problem of inbreeding is typically solved so long as all the adolescents of one sex go and make their fortune elsewhere; the members of the other sex can remain at home and mate with the newcomers immigrating to the group. In chimpanzees and gorillas, it is typically the females who leave for new groups, while the males stay home with their mothers. But among most Old World monkeys—baboons, macaques, langurs—it is the males who make the transfer. Why one sex transfers in one species but not in another is a complete mystery, the sort that keeps primatologists happily arguing with each other ad nauseam.

This pattern of adolescent transfer solves the specter of inbreeding. To look at it just as a solution to an evolutionary problem is mechanistic, however. You can't lose sight of the fact that those are real animals going through the harrowing process. It is a remarkable thing to observe: every day, in the world of primates, someone young and frightened picks up, leaves Mommy and everyone it knows, and heads off into the unknown.

The baboons that I study each summer in East Africa provide one of the best examples of this pattern of adolescent transfer. Two troops encounter each other at midday at some sort of natural boundary—a river, for example. As is the baboon propensity in such settings, the males of the two troops carry on with a variety of aggressive displays, hooting and hollering with what they no doubt hope is a great air of menace. Eventually everyone gets bored and goes back to eating and lounging, ignoring the interlopers on the other side of the river. Suddenly you spot the kid—some adolescent in your troop. He stands there at the river's edge, absolutely riveted. New baboons, a whole bunch of 'em! He runs five steps toward them, runs four back, searches among the other members of his troop to see why no one else seems mesmerized by the strangers. After endless contemplation, he gingerly crosses the river and sits on the very edge of the other bank, scampering back down in a panic should any new baboon so much as glance at him.

A week later. When the troops run into each other again, the kid repeats the pattern. Except this time he spends the afternoon sitting at the edge of the new troop. At the next encounter he follows them for a short distance before the anxiety becomes too much and he turns back. Finally, one brave night, he stays with them. He may vacillate awhile longer, perhaps even ultimately settling on a third troop, but he has begun his transfer into adulthood.

And what an awful experience it is—a painfully lonely, peripheralized stage of life. There are no freshman orientation weeks, no cohorts of newcomers banding together and covering their nervousness with bravado. There is just a baboon kid, all alone on the edge of a new group, and no one there could care less about him. Actually, that is not true—there are often members of the new troop who pay quite a lot of attention, displaying some of the least-charming behavior seen among social primates, and most reminiscent of that of their human cousins. Suppose you are a low-ranking member of that troop, a puny kid a year or so after your own transfer. You spend most of your time losing fights, being pushed around, having food ripped off from you by someone of higher rank. You have a list of grievances a mile long and there's little you can do about it. Sure, there are youngsters in the troop that you could harass pretty successfully, but if they are pretransfer age, their mothers—and maybe their fathers and their whole extended family—will descend on you like a ton of bricks. Then suddenly, like a gift from heaven, a new, even punier kid shows up: someone to take it out on. (Among chimps, where females do the transferring, the same thing occurs; resident females are brutally aggressive toward the new female living on the group's edge.)

Yet that's only the beginning of a transfer animal's problems. When, as part of my studies of disease patterns among baboons, I anesthetize and examine transfer males, I find that these young animals are teeming with parasites. There is no longer anyone to groom them, to sit with them and methodically clean their fur, half for hygiene, half for friendship. And if no one is interested in grooming a recent transfer animal, certainly no one is interested in anything more intimate than that—it's a time of life where sex is mostly just practicing. The young males suffer all the indignities of being certified primate nerds.

They are also highly vulnerable. If a predator attacks, the transfer animal—who is typically peripheral and exposed to begin with—is not likely to recognize the group's signals and has no one to count on for his defense. I witnessed an incident like this once. The unfortunate animal was so new to my group that he rated only a number, 273, instead of a name. The troop was meandering in the midday heat and descended down the bank of a dry streambed into some bad luck: a half-asleep lioness. Panic ensued, with animals scattering every which way as the lioness stirred—while Male 273 stood bewildered and terribly visible. He was badly mauled and, in a poignant act, crawled for miles to return to his former home troop to die near his mother.

In short, the transfer period is one of the most dangerous and miserable times in a primate's life. Yet, almost inconceivably, life gets better. One day an adolescent female will sit beside our nervous transfer male and briefly groom him. Some afternoon everyone hungrily descends on a tree in fruit and the older adolescent males forget to chase the newcomer away. One morning the adolescent and an adult male exchange greetings (which, among male baboons, consists of yanking on each other's penis, a social gesture predicated on trust if ever I saw one). And someday, inevitably, a terrified new transfer male appears on the scene, and our hero, to his perpetual shame perhaps, indulges in the aggressive pleasures of finding someone lower on the ladder to bully.

In a gradual process of assimilation, the transfer animal makes a friend, finds an ally, mates, and rises in the hierarchy of the troop. It can take years. Which is why Hobbes was such an extraordinary beast.

Three years ago my wife and I spent the summer working with a baboon troop at Amboseli National Park in Kenya, a research site run by the behavioral biologists Jeanne and Stuart Altmann from the University of Chicago. When I'm in the field, I study the relationships between a baboon's rank and personality, how its body responds to stress, and what sorts of stress-related diseases it acquires. In order to obtain the physiological data—blood samples to gauge an animal's stress hormone levels, immune system function, and so on—you have to anesthetize the baboon for a few hours with the aid of a small aluminum blowgun and drug-filled darts. Fill the dart with the right amount of anesthetic, walk up to a baboon, aim, and blow, and he is snoozing five minutes later. Naturally it's not quite that simple—you can only dart someone in the mornings (to control for the effect of circadian rhythms on the animal's hormones). You must ensure there are no predators around to shred the guy, and you must make certain he doesn't climb a tree before passing out. And most of all, you have to dart and remove him from the troop when none of the other baboons are looking so that you don't alarm them and disrupt their habituation to the scientists. So essentially, what I do with my college education is creep around in the bushes after a bunch of baboons, waiting for the instant when they are all looking the other way so that I can zip a dart into someone's tush.

It was about halfway through the season. We were just beginning to know the baboons in our troop (named Hook's troop, for a long-deceased matriarch) and were becoming familiar with their daily routine. Each night they slept in their favorite grove of trees; each morning they rolled out of bed to forage for food in an open savannah strewn with volcanic rocks tossed up aeons ago by nearby Mount Kilimanjaro. It was the dry season, which meant the baboons had to do a bit more walking than usual to find food and water, but there was still plenty of both, and the troop had time to lounge around in the shade during the afternoon heat. Dominating the social hierarchy among the males was an imposing character named Ruto, who had joined the troop a few years before and had risen relatively quickly in the ranks. Number two was a male named Fatso, who had had the misfortune of being a rotund adolescent when he'd transferred into the troop years before and was named by a callous researcher. Fat he was no longer. Now a muscular prime-age male, he was Ruto's most obvious competitor, though still clearly subordinate. It was a fairly peaceful period for the troop; mail was delivered regularly and the trains ran on time.

Then one morning we arrived to find the baboons in complete turmoil. It is not an anthropomorphism to say that everyone was mightily frazzled. There was a new transfer male, and not someone meekly scurrying about the periphery. He was in the middle of the troop, raising hell—threatening and chasing everyone in sight. This nasty, brutish animal was soon named Hobbes (in deference to the seventeenth-century English philosopher who described the life of man as solitary, poor, nasty, brutish, and short).

This is an extremely rare, though not unheard of, event among baboons. The transfer male in such cases is usually a big, muscular, intimidating kid. Maybe he is older than the average seven-year-old transfer male, or perhaps this is his second transfer and he picked up confidence from his first emigration. Maybe he is the cocky son of a high-ranking female in his old troop. In any case, the rare animal with these traits comes on like a locomotive, and he often gets away with it for a time. As long as he keeps pushing, it will be a while before any of the resident males works up the nerve to confront him. Nobody knows him yet; thus no one knows if he is asking for a fatal injury by being the first fool to challenge this aggressive maniac.

This was Hobbes's style. The intimidated males stood around helplessly. Fatso discovered all sorts of errands he had to run elsewhere. Ruto hid behind females. No one else was going to take a stand. Hobbes rose to the number one rank in the troop within a week.

Despite his sudden ascendancy, Hobbes's success wasn't going to last forever. He was still a relatively inexperienced kid, and eventually one of the bigger males was bound to cut him down to size. Hobbes had about a month's free ride. At this point he did something brutally violent but which had a certain grim evolutionary logic. He began to selectively attack pregnant females. He beat and mauled them, causing three out of four to abort within a few days.

One of the great clichés of animal behavior in the context of evolution is that animals act for the good of the species. This idea was discredited in the 1960s but continues to permeate *Wild Kingdom*–like versions of animal behavior. The more accurate view is that animals usually behave in ways that maximize their own reproduction and the reproduction of their close relatives. This helps explain extraordinary acts of altruism and self-sacrifice in some circumstances, and sickening aggression in others. It's in this context that Hobbes's attacks make sense. Were those females to carry through their pregnancies and raise their offspring, they would not be likely to mate again for two years—and who knows where Hobbes would be at that point. Instead he harassed the females into aborting, and they were ovulating again a few weeks later. Although female baboons have a say in whom they mate with, in the case of someone as forceful as Hobbes they have little choice: within weeks Hobbes, still the dominant male in the troop, was mating with two of those three females. (This is not to imply that Hobbes had read his textbooks on evolution, animal behavior, and primate obstetrics and had thought through this strategy, any more than animals think through the logic of when they should reach puberty. The wording here is a convenient shorthand for the more correct way of stating that his pattern of behavior was almost certainly an unconscious, evolved one.)

As it happened, Hobbes's arrival on the scene—just as we had darted and tested about half the animals for our studies—afforded us a rare opportunity. We could compare the physiology of the troop before and after his tumultuous transfer. And in a study published with Jeanne Altmann and Susan Alberts, I documented the not very surprising fact that Hobbes was stressing the bejesus out of these animals. Their blood levels of cortisol (also known as hydrocortisone), one of the hormones most reliably secreted during stress, rose significantly. At the same time, their numbers of white blood cells, or lymphocytes, the sentinel cells of the immune system that defend the body against infections, declined markedly, another highly reliable index of stress. These stress-response markers were most pronounced in the animals receiving the most grief from Hobbes. Unmolested females had three times as many circulating lymphocytes as one poor female who was attacked five times during those first two weeks.

An obvious question: Why doesn't every new transfer male try something as audaciously successful as Hobbes's method? For one thing, most transfer males are too small at the typical transfer age of seven years to intimidate a gazelle, let alone an eighty-pound adult male baboon. (Hobbes, unusually, weighed a good seventy pounds.) Most don't have the personality needed for this sort of unpleasantry. Moreover, it's a risky strategy, as someone like Hobbes stands a good chance of sustaining a crippling injury early in life.

But there was another reason as well, which didn't become apparent until later. One morning, when Hobbes was concentrating on whom to hassle next and paying no attention to us, I managed to put a dart into his haunches. Months later, when examining his blood sample in the laboratory, we found that Hobbes had among the highest levels of cortisol in the troop and extremely low lymphocyte counts, less than one-quarter the troop average. (Ruto and Fatso, conveniently sitting on the sidelines, had three and six times as many lymphocytes as Hobbes had.) The young baboon was experiencing a massive stress response himself, larger even than those of the females he was traumatizing, and certainly larger than is typical of the other, meek transfer males I've studied. In other words, it doesn't come cheap to be a bastard twelve hours a day—a couple of months of this sort of thing is likely to exert a physiological toll.

As a postscript, Hobbes did not hold on to his position. Within five months he was toppled, dropping down to number three in the hierarchy. After three years in the troop, he disappeared into the sunset, transferring out to parts unknown to try his luck in some other troop.

All this only reaffirms that transferring is awful for adolescents—whether they're average geeks opting for the route of slow acceptance or rare animals like Hobbes who try to take a troop by storm. Either way it's an ordeal, and the young animals pay a heavy price. The marvel is that they keep on doing it. Transferring may solve the inbreeding problem for a population, but what's in it for the individual?

Adolescent transfer is a feature of many social mammals, not just primates, and the mechanisms can differ. Sometimes transfer can arise from intrasexual competition—a fancy way of saying that the adolescents are driven from the group by a more powerful same-sex competitor. You usually see this in species like gazelles and impalas. The core social group consists of a single breeding male, a large collection of females, and their offspring. At any given point some of those male offspring are likely to be entering puberty. But since the breeding male typically doesn't hold on to his precarious position for long, he is probably not the father of these adolescent males. He doesn't view them as sons coming of age but as unrelated males growing into reproductive competitors. At their first signs of puberty he violently drives them out of the group.

In primates, however, forced dispersion almost never happens. Something absolutely remarkable occurs: these adolescents *choose* to go—even though the move seems crazy. After all, they live in a troop surrounded by family and friends. They know their home turf, which trees are fruiting at what time of year, where the local predators lurk. Yet they leave these home comforts to endure parasites, predators, and loneliness. And why? To dwell among strangers who treat them terribly. It makes no sense, from the standpoint of the individual animal. Behaviorist theories state this more formally: Animals, including humans, tend to do things for which they are rewarded and tend not to do things for which they get punished. Yet here they are, leaving their comfortable, rewarding world in order to be amply dumped on far away. Furthermore, animals usually hate novelty. Place a rat in a new cage, give it a new feeding pattern, and it exhibits a stress response. Yet here young primates are risking life and limb for novelty. Old World monkeys have been known to transfer up to five times over their lifetime, or travel nearly forty miles to a new troop.

Why should any individual in its right mind want to do this?

I do not know why transfer occurs, but it is clearly very deeply rooted. Humans, in part because their diets are adaptable, are the most widely distributed mammal on earth, inhabiting nearly every godforsaken corner of this planet. Among our primate relatives, those with the least finicky of diets, such as baboons, are also among the most widely distributed beasts on earth (African baboons range from desert to rain forest, from mountain to savannah). Inevitably, someone had to be the first to set foot in each of those new worlds, an individual who transferred in a big way. And it is overwhelmingly likely it was a young individual who did that.

This love affair with risk and novelty seems to be why the young of all our primate species are the most likely to die of accidents, doing foolhardy things while their elders cluck over how they told them so.* And it is also the reason that the young are most likely to discover something really new and extraordinary, whether in the physical or the intellectual realm. When the novel practice of washing food in seawater was discovered by snow monkeys in Japan, it was a youngster who did so, and it was her playmates who picked up the adaptation; hardly any of the older animals did. And when Darwin's ideas about evolution swept through academic primates in the mid-nineteenth century, it was the new, up-and-coming generation of scientists that embraced his ideas with the greatest enthusiasm.

You don't have to search far for other examples. Think of the tradition of near-adolescent mathematicians revolutionizing their fields, or of the young Picasso and Stravinsky inventing twentieth-century culture. Think about teenagers inevitably, irresistibly wanting to drive too fast, or trying out

---

*I discovered a wonderful example of this recently. In the Sierra foothills there are the California Caverns, a gigantic cave system that begins, after an initial narrow twisting descent of 30-some feet, with an abrupt 180-foot drop (now navigable with either a staircase or a tourist-friendly rappel). The Park Service found scores of ancient human skeletons at the bottom of that drop, dating back centuries, millennia. Knowledge of the Native American cultures living there at those times argued strongly against these being human sacrifices. Instead, they were explorers who took one step too far in the gloom. And the skeletons are always of adolescents.

some new improvisatory sport guaranteed to break their necks, or marching off in an excited frenzy to whatever stupid war their elders have cooked up. Think of the endless young people leaving their homes, homes perhaps rife with poverty or oppression, but still their homes, to go off to new worlds.

Part of the reason for the evolutionary success of primates, human or otherwise, is that we are a pretty smart collection of animals. What's more, our thumbs work in particularly fancy and advantageous ways, and we're more flexible about food than most. But our primate essence is more than just abstract reasoning, dexterous thumbs, and omnivorous diets. Another key to our success must have something to do with this voluntary transfer process, this primate legacy of feeling an itch around adolescence. How did voluntary dispersal evolve? What is going on with that individual's genes, hormones, and neurotransmitters to make it hit the road? We don't know, but we do know that following this urge is one of the most resonantly primate of acts. A young male baboon stands riveted at the river's edge; an adolescent female chimp cranes to catch a glimpse of the chimps from the next valley. New animals, a whole bunch of 'em! To hell with logic and sensible behavior, to hell with tradition and respecting your elders, to hell with this drab little town, and to hell with that knot of fear in your stomach. Curiosity, excitement, adventure—the hunger for novelty is something fundamentally daft, rash, and enriching that we share with our whole taxonomic order.

FURTHER READING

For a good general overview of intertroop transfer among primates, see A. Pusey and C. Packer, "Dispersal and Philopatry," in B. Smuts, D. Cheney, R. Seyfarth, R. Wrangham, and T. Struhsackers, eds., *Primate Societies* (Chicago: University of Chicago Press, 1987), 250. For a very detailed analysis of the costs and benefits of transfer among baboons, see S. Alberts and J. Altmann, "Balancing Costs and Opportunities: Dispersal in Male Baboons," *American Naturalist* 145 (1995): 279.

For the story of Hobbes, in dauntingly technical detail, see S. Alberts, J. Altmann, and R. Sapolsky, "Behavioral, Endocrine and Immunological Correlates of Immigration by an Aggressive Male into a Natural Primate Group," *Hormones and Behavior* 26 (1992): 167.

The Solace of Patterns

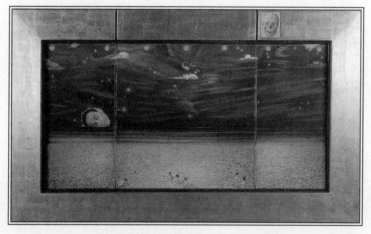

Juan Gonzales, Mar de Lagrimas, *1987–88;*
courtesy Nancy Hoffman Gallery, New York. Photo: Chris Watson

A short time ago my father died, having spent far too many of his last years in pain and degeneration. Although I had expected his death and tried to prepare myself for it, when the time came it naturally turned out that you really can't prepare. A week afterward I found myself back at work, bludgeoned by emotions that swirled around a numb core of unreality—a feeling of disconnection from the events that had just taken place on the other side of the continent, of disbelief that it was really him frozen in that nightmare of stillness. The members of my laboratory were solicitous. One, a medical student, asked me how I was doing, and I replied, "Well, today it seems as if I must have imagined it all." "That makes sense," she said. "Don't forget about DABDA."

DABDA. In 1969 the psychiatrist Elisabeth Kübler-Ross published a landmark book, *On Death and Dying.* Drawing on her research with terminally ill people and their families, she described the process whereby people mourn the death of others and, when impending, of themselves. Most of us, she observed, go through a fairly well-defined sequence of stages. First we deny the death is happening. Then we become angry at the unfairness of it all. We pass through a stage of irrational bargaining, with the doctors, with God: Just let this not be fatal and I will change my ways. Please, just wait until Christmas. There follows a stage of depression and, if one is fortunate, the final chapter, serene acceptance. The sequence is not ironclad; individuals may skip certain stages, experience them out of order, or regress to earlier ones. DABDA, moreover, is generally thought to give a better description of one's own preparation for dying than of one's mourning the demise of someone else. Nevertheless, there is a broadly recognized consistency in the overall pattern of mourning: denial, anger, bargaining, depression, acceptance. I was stuck at stage one, right on schedule.

Brevity is the soul of DABDA. A few years ago I saw that point brilliantly dramatized on television—on, of all programs, *The Simpsons.* It was the episode in which Homer, the father, accidentally eats a poisonous fish and is told he has twenty-four hours to live. There ensues a thirty-second sequence in which the cartoon character races through the death and dying stages, something like this: "No way! I'm not dying." He ponders a second, then grabs the doctor by the neck. "Why, you little . . ." He trembles in fear, then pleads, "Doc, get me outta this! I'll make it worth your while." Finally he composes himself and says, "Well, we all gotta go sometime." I thought it was hilarious. Here was a cartoon suitable to be watched by kids, and the writers had sneaked in a parody of Kübler-Ross.

For sheer conciseness, of course, Homer Simpson's vignette has nothing on the phrase itself, "DABDA." That's why medical students memorize the acronym along with hundreds of other mnemonic devices in preparation for their national board examinations. What strikes me now is the power of those letters to encapsulate human experience. My father, by dint of having been human, was unique; thus was my relationship to him, and thus must be my grieving. And yet I come up with something reducible to a medical school acronym. Poems, paintings, symphonies by the most creative artists who ever lived, have been born out of mourning; yet, on some level, they all sprang from the pattern invoked by two pedestrian syllables of pseudo-English. We cry, we rage, we demand that the oceans' waves stop, that the planets halt their movements in the sky, all because the earth will no longer be graced by the one who sang lullabies as no one else could; yet that, too, is reducible to DABDA. Why should grief be so stereotypical?

Scientists who study human thought and behavior have discerned many stereotyped, structured stages through which all of us move at various times. The logic of some of the sequences is obvious. It is no surprise that infants learn to crawl before they take their first tentative steps, and only later learn to run. Other sequences are more subtle. Freudians claim that in normal development the child undergoes the invariant transition from a so-called oral stage to an anal stage to a genital stage, and they attribute various aspects of psychological dysfunction in the adult to an earlier failure to move successfully from one stage to the next.

Similarly, the Swiss psychologist Jean Piaget mapped stages of cognitive development. For example, he noted, there is a stage at which children begin to grasp the concept of object permanence: Before that developmental transition, a toy does not exist once it is removed from the child's sight. Afterward, the toy exists—and the child will look for it— even when it is no longer visible. Only at a reliably later stage do children begin to grasp concepts such as the conservation of volume—that two pitchers of different shapes can hold the same quantity of liquid. The same developmental patterns occur across numerous cultures, and so the sequence seems to describe the universal way that human beings learn to comprehend a cognitively complex world.

The American psychologist Lawrence Kohlberg mapped the stereotyped stages people undergo in developing morally. At one early stage of life, moral decisions are based on rules and on the motivation to avoid punishment: actions considered for their effects on oneself. Only at a later stage are decisions made on the basis of a respect for the community, where actions are considered for their effects on others. Later still, and far more rarely, some people develop a morality driven by a set of their own internalized standards derived from a sense of what is right and what is wrong for all possible communities. The pattern is progressive: People who now act out of conscience invariably, at some earlier stage of life, believed that you don't do bad things because you might get caught.

The American psychoanalyst Erik Erikson discerned a sequence of psychosocial development, framing it as crises that a person resolves or fails to resolve at each stage. For infants, the issue is whether one attains a basic attitude of trust toward the world; for adolescents, it is identity versus identity confusion; for young adults, intimacy versus isolation; for adults, generativity versus stagnation; and for the aged, peaceful acceptance and integrity versus despair. Erikson's pioneering insight that one's later years represent a series of transitions that must be successfully negotiated is reflected in a quip by the geriatrician Walter M. Bortz II of Stanford University Medical School. Asked whether he was interested in curing aging, Bortz responded, "No, I'm not interested in arrested development."

Those are some of the patterns we all are reported or theorized to have in common, across many settings and cultures. I think such conceptualizations are often legitimate, not just artificial structures that scientists impose on inchoate reality. Why should we share such patterning? It is certainly not for lack of alternatives. As living beings, we represent complex, organized systems—an eddy in the random entropy of the universe. When all the possibilities are taken into account, it is supremely unlikely for elements to assemble themselves into molecules, for molecules to form cells, for vast assemblages of cells to form us. How much more unlikely, it seems, that such complex organisms conform to such relatively simple patterns of behavior, of development, or of thought.

One way of coming to grips with the properties of complex systems is through a field of mathematics devoted to the study of so-called cellular automata. The best way of explaining its style of analysis is by example. Imagine a long row of boxes—some black, some white—arranged to form some initial pattern, a starting stage. The row of boxes is to give rise to a second row, just below the first. The way that takes place in a cellular automation is that each box in the first row is subjected to a set of reproduction rules. For example, one rule might stipulate that a black box in the first row gives rise to a black box immediately below it in the next row only if exactly one of its two nearest neighbors is black. Other rules might apply to a black box flanked by two white boxes or two black boxes. Once the set of rules is applied to each box in the first row, a second row of black and white boxes is generated; then the rules are applied again to each box in the second row to generate a third row and so on.

Metaphorically, each row represents one generation, one tick of a clock. A properly programmed computer could track any possible combination of colored boxes, following any conceivable set of reproduction rules, down through the generations. In the vast majority of cases, somewhere down the line it would end up with a row of boxes all the same color. After that, the single color would repeat itself forever. In other words, the line would go extinct.

Return now to my earlier question: How can it be, in this entropic world, that we human beings share so many stable patterns—one nose; two eyes; a reliable lag time before we learn object permanence; happier adulthoods if we become confident about our identities as adolescents; a tendency to find it hard to believe in tragedy when it strikes. What keeps us from following an almost infinite number of alternative developmental paths? The studies of cellular automata provide a hint.

Not all complex patterns, it turns out, eventually collapse into extinction. A few combinations of starting states and reproduction rules beat the odds and settle down into mature stable patterns that continue down through the generations forever. In general, it is impossible to predict whether a given starting state will survive, let alone which pattern it will generate after some large number of generations. The only way to tell is to crank it through the computer and see. And when you do this, there is a surprise—only a small number of such mature patterns are possible.

A similar tendency in living systems has long been known to evolutionary biologists. They call it convergence. Among the staggering number of species on this planet, there are only a few handfuls of solutions to the problem of how to locomote, how to conserve fluids in a hot environment, how to store and mobilize energy. And among the staggering variety of humans, it may be a convergent feature of our complexity that there are a small number of ways in which we grow through life or mourn its inevitabilities.

In an entropic world, we can take a common comfort from our common patterns, and there is often consolation in attributing such patterns to forces larger than ourselves. As an atheist, I have long taken an almost religious solace from a story by the Argentine minimalist Jorge Luis Borges. In his famous short story "The Library of Babel," Borges describes the world as a library filled with an unimaginably vast number of books, each with the same number of pages and the same number of letters on each page. The library contains a single copy of every possible book, every possible permutation of letters. People spend their lives swimming through this ocean of gibberish, searching for the incalculably rare books whose random arrays of letters form something meaningful, searching above all else for the single book (which must exist) that explains everything, the book that gives the history of all that came before and all that will come, the book containing the dates of birth and of death of every human who would ever wander the halls of the library. And of course, given the completeness of the library, in addition to that perfect book, there must also be one that convincingly disproves the conclusions put forth in it, and yet another book that refutes the malicious solipsisms of the second book . . . plus hundreds of thousands of books that differ from any of those three by a single letter or a comma.

The narrator writes in his old age, in an isolation brought about by the suicides of people who have been driven to despair by the futility of wandering through the library. In this parable of the search for meaning amid entropy, Borges concludes:

> *Those who judge {the library to be finite} postulate that in remote places the corridors and stairways and hexagons can conceivably come to an end—which is absurd. Those who imagine it to be without limit forget that the possible number of books does have such a limit. I ventured to suggest this solution to the ancient problem: The library is unlimited and cyclical. If an eternal traveler were to cross it in any direction after centuries he would see that the same volumes were repeated in the same disorder (which, thus repeated, would be an order: the Order). My solitude is gladdened by this elegant hope.*

It appears that amid the order with which we mature and decline, there is an order to our mourning. And my own recent solitude is gladdened by that elegant hope, in at least two ways. One is inward-looking. This stereotypy, this ordering, brings the promise of solace in the predicted final stage: If one is fortunate, DABDA ends in A.

Another hope looks outward, to a world whose tragedies are inexorably delivered from its remotest corners to our nightly news. Look at the image of a survivor of some carnage and, knowing nothing of her language, culture, beliefs, or circumstances, you can still recognize in the fixed action patterns of her facial muscles the unmistakable lineaments of grief. That instant recognition, the universal predictability of certain aspects of human beings, whether in a facial expression or in the stages of mourning, is an emblem of our kinship and an imperative of empathy.

FURTHER READING

Readers may be interested in reading the classic E. Kübler-Ross, *On Death and Dying* (New York: Macmillan, 1969).

A good introduction to cellular automata can be found in G. Ermentrout and L. Edelstein-Keshet, "Cellular Automata Approaches to Biological Modeling," *Journal of Theoretical Biology* 160 (1993): 97. Be forewarned: This is not bedtime reading.

Borges's "Library of Babel" can be found in J. Borges, *Labyrinths: Selected Stories and Other Writings* (New York: New Directions Books, 1962).

Beelzebub's SAT Scores

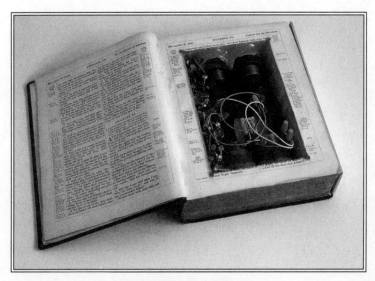

Gregory Green, Bible Bomb, 1996;
courtesy Max Protetch Gallery, New York

If this piece were to be published in a magazine like *Cosmo* or *GQ,* I'd bet that very few readers would have a glimmer of what I will be talking about. And because of the sort of book this is and the readers it attracts, I think my subject is going to strike an uncomfortable, responsive note among far more of you than you'd like to admit.

You don't have to be brilliant to be interested in science, but it does require certain values. Facts are deemed more important than intuition or rhetoric or sound bites when it comes to making certain types of decisions, as are statistics over a good but singular anecdote. Solving a sustained, complex problem, one even stretching on for years, can be more satisfying than the viscera of whatever phenomenon is having its fifteen minutes of fame. And the products of the mind can not only be vital to master for some utilitarian reason, but can also be valued because they bring an elegant, beautiful joy.

Nothing comes for nothing. A fondness for thinking often carries a cost. Sure, everyone had their adolescent miseries, but there is a characteristic type for the ones who were eggheads. This is the adolescent world of being picked last for sports teams or, worse, being picked first when some displacement aggression was in order. A world of always being a suitable source of answers to homework questions but never a suitable date. The ludicrous array of deflective personality twitches meant to cover up the fact that you were smart and into school or *really* obsessed with hermit crabs or topology or plate tectonics. And even decades later, even among those who metamorphed into happy, fulfilled, secure adults, there is often still a necrotic core of anger somewhere down there at how bad it was back when. These are the stigmata of geekdom.

As with any culture, geekdom comes with its deities. There was the old reliable pantheon, Einstein occupying the chair of Zeus, the worship of my parents' generation of Eleanor Roosevelt and Adlai Stevenson for being smart as well as compassionate in an arena that rarely rewarded either. Before he fell from the heavens by making serious movies and serious moves on his nineteen-year-old pseudodaughter, there was the incomparable pleasure of Woody Allen. He epitomized every trait for which we were all once punished, yet in the movie world that he created, it didn't matter; it even worked to his advantage—he always got the goil. And, more wonderfully, he did so in real life. For a while, there was the ascendancy of Michael Dukakis, a ripple of recognition and admiration felt for him. He hit all those themes with a vengeance—the smart geek and, by dint of his first-generation ethnicity, even more the perpetually awkward outsider. Stiff, ungraceful, humorless, excelling in a world of privileged style through sheer smarts and discipline and sincerity. And for an instant, he might even have become president. And, to complete the Greek tragedy that we all identified with, he was brought low by the fact that ultimately, all he had going for him was smarts and discipline and sincerity, done in by a man who, with the self-satisfied calm of a ruling-class Brahmin, played dirty with a Willie Horton up his sleeve. That was most certainly a case where we felt his pain.

For every yin there is a yang, and the gods on high must have their counterparts from the Dark Side. For a community of intellectuals, these have traditionally been the outsiders who attack the values we cherish, the yahoos and know-nothings who would vote for guns over books, who value ideologues rather than thought. That we are accustomed to. What is difficult for us (and for any community) is when the menace comes from within, because of the special challenge it presents. And that is what has happened.

In recent months, plying the Internet, talking to friends and colleagues, gauging the reactions of students to throwaway lines in lectures, it has become apparent to me that there is a furtive, guilty, confused excitement in the community of learned individuals, a fascination that no one knows quite what to do with: the intelligentsia have found their Antichrist, and he is Ted.

There seems to be an overlapping array of facets that people are focusing on. Certainly, some are responding to Kaczynski because of his obsessional grudges, feel a resonance with the political and social undertones of his antitechnology mania (and I've noted a lot of right-wing media picking up on this theme, as if somehow Berkeley in the sixties were responsible). But most don't seem to care about Kaczynski's Luddism. Instead, it's easy to indulge in some cheap psychoanalyzing, deciding that he wouldn't have cared as much about technology and society if he had gotten laid more often.

Some seem to be most fascinated with his chosen route of (required by law: alleged) murderousness. With the sickening regularity of drive-by shootings or someone going postal, or even with the rarity of a rental van crammed with tons of explosive fertilizer, this time, tiny, handcrafted objects were placed inside an envelope and, sometime later across a continent, someone was shredded. It's as if the link between cause and effect were weakened. Something about the detachment and precision fascinates. We are accustomed to the discover-the-spouse-in-bed-with-the-best-friend-and-out-comes-the-gun crime of passion: it seems like the consequence of an inflamed limbic system, the part of the brain involved in emotion. Instead, with Kaczynski, we feel as if we are seeing the consequences of an inflamed cortex.

But most people are not responding to his content or method. Instead, it is the persona, the smarts coupled with the social disconnection, the crazy hermit life in that cabin filled with Shakespeare and foreign literature and bomb manuals.

On the more benign side, that persona pulls among the intelligentsia because of its surprising familiarity. I did my schooling at the proud institution that produced Ted K. Following his arrest, I played a game with my fellow alums— how many people from your dorm could have turned out to be the Unabomber? Maybe because we were all the insecure, driven ethnics with something to prove, instead of the umpteenth-generation Harvardians drinking at their club and gliding through with gentlemen's C's, we seem to have no shortage of suspects: the fifteen-year-old physicist who never said a word and who paced his room at night, pulling out his hair. The political science major with the foul rotting breath who drove everyone from the dinner table with her trembling, tearful rantings about her latest obsession. The mathematician with no friends who, as part of some odd social experiment, would go live on the streets, subsisting off garbage Dumpsters, emerging for brief periods after longer and longer stretches. The smarts and the disconnection and the simmering obscure anger—but for the grace of God plus some fine-motor control when it came to handling small tools, any of them could have been him.

So part of it is the recognition, the unexpected familiarity. There is a slight class snootiness to this, explaining one facet of the odd fascination I'm detecting. It reminds me of the quip that the McCarthy era consisted of a bunch of Fordham law school graduates trying to get the goods on a bunch of Harvard law school graduates. Typically, whenever some violent criminal of reptilian coldness emerges into the nighttime news, there is the predictable alienness for this intellectual community—the childhood trailer park mired in poverty and tornadoes, the drunken abandoning father, the reform schools, the bad teeth and bad grammar and sweat rings around the armpits, the trail of bodies hacked up and dumped from a rusty pickup truck into the bayou. How déclassé. At last there is a serial murderer we can call our own.

But there is an even more disturbing component, the one that seems hardest for people to admit to but which I am detecting again and again in offhanded, self-conscious quips: rather than mere recognition, there is a frightening hint of identification, of there but for the grace of God go I. We all have our dark sides, a world of humid, spittle-flecked, unrestrained impulse that goes on in our heads. While it is obligatory to trash Freud as being a repressive dead white male, it is useful on occasion to instead remember him as a great liberator, the man who tried to break the back of centuries of authoritarian, religious judgment about those internal worlds, the man who discovered the unconscious and thus taught us that thought is no crime. We all do indeed have our dark sides. One evening, that great horned toad of an awkward intellectual, Karl Marx, came home from fulminating in the British Museum. "At any rate," he wrote to Engels that night, "I hope the bourgeoisie will remember my carbuncles all the rest of their lives." As well they did. Few of us can ever hope for that level of retributive pissiness. We merely fantasize about returning someday to our childhood neighborhoods, encountering the ex-bullies or the catty girls who were in the in-group when we were not, and beating them into contused, bloodied contrition with our thick stack of diplomas. Actually, our dark fantasies of anger and revenge would rarely run in the Rambo action direction. Instead, we all have our little lists of enemies to be vanquished with the force of our minds, in clever, cerebral, elegant bits of mayhem. And then Kaczynski goes and does something with some queasy, identificative elements of exactly that. And worst of all, our pulses quicken ever so slightly, as if, sickeningly, we are almost proud: take thousands of big strapping guys who no doubt all once got to go to the prom with homecoming queens, give 'em guns and badges and FBI good looks, and it still takes them eighteen years to catch weird smart Ted with his crazy hair. Ted Kaczynski, the Id from Mensa.

Thus, this fascination arising from the reflexive respect we have for intelligence, from the detachment of his methods, from his resemblance to the loner down the freshman hallways, his resemblance to the It inside us. There is the danger of a certain empathy creep, the transition from recognition to understanding and then to something resembling forgiveness. And thus, the remainder of this piece must be about the reassertion of our superegos.

There is another arena in which the issue of understanding makes it hard for thoughtful, educated individuals in the criminal justice system to mete out judgment. Epileptics who in the course of a seizure flail their arms and strike someone are, of course, not assaultive. Their arms may swing violently, but they are not violent people. "It's not him, it's his disease." We are coming to recognize a shocking number of ways in which biology can impact and distort the essence of who a person is—certain hormonal or neurotransmitter imbalances, tumors in parts of the limbic system, damage to the frontal cortex—producing someone who is uncontrollably destructive. In this arena, I don't think one has to steel oneself to avoid the transition from understanding to forgiveness, because "forgiveness" becomes as irrelevant as branding as "evil" a car whose brakes are cut and careens destructively. You fix the car, if possible, but you mostly find ways to protect society from such cars. Medicalizing people into being broken cars is dehumanizing, but still a hell of a lot more humane than moralizing them into being sinners.

But this is not the problem with Kaczynski; the inflammation of his cortex is just a metaphor. There appears, at this date, to be no evidence of irresistible voices in his ears, of a childhood head injury that damaged components of his nervous system that inhibit impulsivity or aggressiveness. There is no evidence of a biological imperative that cut his brakes. The reasons are different why we have to stand fast in the face of our fascination with him.

The issue of his methods is easily disposed of, especially for those of us who once protested the detachment of a button being pressed in a high-altitude bomber and, as a result, a Vietnamese village being obliterated. Our era is dominated by violence that involves pushing a button or giving an order or looking the other way, rather than by hands with barely opposing thumbs gripping cudgels. In that context, mailing a certain kind of letter is just another twentieth-century act of violence. But if technology has lengthened the reach of violence, it has also lengthened the range of acts we can be sickened by and must act against.

The next reason why we have to stand firm is related to why we jailed Ezra Pound, the problem of what to do when good poets do bad things. The Kaczynski conundrum is actually more watered-down than that. Kaczynski did not savage people while, nonetheless, producing great works (just check out his Manifesto in the john tonight if you don't believe me). There is simply his ability to appreciate great works. Its equivalence is one that haunts me—the certainty that there was at least one murderous Nazi who, at the end of a difficult day, could sit and be moved to tears by the same instant of Beethoven that undoes me. It is thoroughly obvious that we can feel no softening toward Kaczynski simply because he is smart or learned and we respect the world of intellect. This is a hard one, part of what has made him the Antichrist to our community and values. Standing fast on this issue requires a heartbreaking rejection of a tenet of liberal education that so many of us treasure. It is the notion that being exposed to the Great Books and the Great Thoughts must lead to Great Morals; unfortunately, we should probably be satisfied if it leads to a decent vocabulary now and then.

But we are most challenged by his being the misfit from down the hall, from down inside of us. While the danger of empathy creep there is considerable, there is one overriding fact, so easy to forget in our fascination, that must destroy it. No matter how familiar Kaczynski is, no matter how often we have had fleeting thoughts like that, no matter how much we think we understand where he came from, there is an infinitely yawning chasm, something that makes him as unfamiliar as a silicon-based life-form, something that makes it clear that we actually understand nothing whatsoever about him or what went on in his head—he actually *did* these things. Thought is no crime, but crime is, and bombs have exploded not as abstractions in someone's mind, but in the faces of innocent people.

There is a wonderful Russian story that takes place at the gates of heaven, where the newly arrived are judged. A dead murderer is on trial, fresh from earth where he was shot by the police after his umpteenth murder, the strangling of an elderly woman for her money. A panel of deceased judges sits in session. And where does God fit on the scene? Not as a judge, but as a required character witness. At some point in the proceedings, he shambles in, sits in a magisterial decrepitude born of the weight of infinite knowledge, and in a meandering, avuncular way, does his best to defend and explain the man—"He was always kind to animals. He was very upset when he lost his favorite top when he was a small boy." ("My red top, you know about my red top!?!" The murderer leaps up, suddenly awash in a torrent of memory. "Of course I do. It rolled down the storm drain on Zlotny Street. It's still down there," God answers with complete, affectless knowing.) Finally, the judges tire of God, who is in fact tiresome in his knowledge and forgiveness, and coax him off the stand.

When science brings us something new and startling, when there is a breakthrough that opens new vistas, there is often talk about us acquiring godlike knowledge, and the tacit assumption is that this is a good thing. But the God of this parable is useless, has been shunted aside by the indiscriminateness of his knowledge. Knowledge, familiarity, understanding, must not ever lead us to a detached indiscriminateness. The danger of Olympian knowledge is that you then look down upon things from an Olympian height, and from that telescoped distance, things seem equivalent—like a lost red top and a strangled woman, or perhaps an awkward adolescence that produces an awkward adult and an awkward adolescence that produces a murderous one.

But there is a difference.

Poverty's Remains

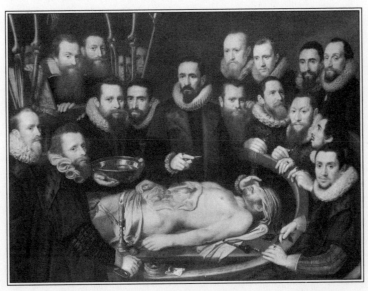

Pieter van Miereveld, Doctor W. Van der Meer's Anatomy Lesson in Delft, *1617;*
Municipal Hospital, Delft, The Netherlands, Giraudon/Art Resource, New York

When the aliens finally visit earth and want to understand human behavior, the first thing they are going to have to sort out is this obsession we have with death. So much of our time is spent simultaneously denying and mourning our evanescent nature, a habit that manifests itself as a profound queasiness whenever we think of ourselves as flesh-and-blood. When I teach neuroanatomy and handle a human brain, pointing out some pathway to students, I never fail to get the willies that the same pathways lie inside my own head. Or try this: the next time a loved one tenderly brushes a hand against your cheek, remember that someday, that cheekbone will be no more than a zygomatic arch in a skull. Or when you're lying in bed tonight, listening to your heart beat, tell yourself, *You know, that heart won't beat forever*—and just wait for the cold sweat to break out.

Such queasiness about one's own flesh projects well into the indefinite future, when, surely, one's body will no longer be living. By all logic, once the battle has been lost, once the inconceivable has actually come to pass, the fate of one's carcass should cease to matter. In 1829, taking such thinking to the extreme, a radical British pamphleteer named Peter Baume left explicit instructions for his body's disposal: his skeleton was to be donated for medical education, or, failing that, his skull left to the phrenological society and his bones made into knife handles and buttons; his skin was to be tanned to make a chair cover, and his soft body parts used as fertilizer for roses.

But most of us don't assume such a pragmatic attitude. We are haunted by thoughts of negligent groundskeepers at cemeteries or, more, by the specter that one day our own burial ground will be paved over for a parking lot. Our graves must be kept clean, our bodies left intact and treated with respect. Armies in the midst of perfect savagery will halt to exchange their dead. In certain religious orthodoxies, body parts removed over the course of a lifetime—teeth, tonsils, a gallbladder—are saved for future burial with the rest of their owner. The mistreatment of the dead can bring outrage, as indigenous populations throughout the world confront anthropology museums and collections and demand the return of the bones of their ancestors. And some extraordinarily poignant moments derive from this concern as well. The thought of a Neanderthal placing flowers around the body of a loved one is tearfully moving; the hand that clutched those flowers may have been less deft than our own, but in that ancient act of grieving we feel the shock of kinship and familiarity.

Such feelings are irrational. What does it matter once we are dead? I have made postmortem arrangements that reflect my belief in the importance of science. Nevertheless, it still unnerves me to think of some callow first-year medical student joking about the appearance of my future, dissected self, or when picturing my beloved spleen pickled in a jar in a dusty corner of some lab.

Who needs that? Few of us, surely; unfortunately, such reticence has traditionally led to a dearth of what anatomists call human biological material. Medical students must learn anatomy; investigators must collect data if they are to distinguish the normal from the diseased; autopsies must be performed so physicians can see what went wrong and thus

enable the living to live longer. But necessary as such proce-
dures are, they are not a pleasing prospect. Most people are
unwilling to donate their bodies, or their organs, to science,
and physicians dread having to persuade next of kin to permit
an autopsy.

Consequently, a disproportionately large fraction of the
cadavers used for instruction and research is made available by
indigent individuals. Because they cannot afford private
physicians, many poor people end their days in large urban
teaching hospitals, where they are studied in sickness and in
death by medical students and interns. But the socioeconom-
ically disadvantaged generally die in ways different from
those of wealthier people, ways that can alter the size and
appearance of their internal organs. As a result anatomists
and medical students, working with this limited and biased
range of material, have acquired a skewed notion of what a
"normal" human body looks like.

The overreliance on the poor for autopsy material dates at
least as far back as the sixteenth century, when King Henry VIII
decreed that the bodies of executed prisoners be turned over to
the anatomists for dissection. The decree was made more for conspicuous punishment than out of a fondness for science; the
name of the cadaver was to be published, the dissection performed publicly, and the remainders tossed to animals. Given
the draconian nature of justice of the time—stealing a piece of
bread could be a capital crime—it was by and large the poor
who wound up on the dissectors' slabs.

Medical schools were growing in size and number, however, and with them the need for anatomical research and
training; even the enthusiasm for public hangings could not
match the demand for cadavers. In the late eighteenth century a new career opportunity was born, the body snatcher.
"Resurrectionists" spent moonless nights digging up fresh
bodies or they might raid a funeral en masse and wrest the

deceased from horrified relatives. Those less inclined to such taxing manual labor might simply bribe the orderly at a hospital death room or pretend to be a relative in order to claim the body.

Again it was overwhelmingly the poor who were purloined. Funerals for the wealthy were guarded; those of the poor were easily raided. The indigent deceased might lie vulnerable in a hospital for days while the family (if one existed) tried to scrape together the funds to pay the hospital and claim the corpse. And the bodies of the poor were buried, at best, in flimsy coffins or, more often, laid coffinless in shallow mass graves or paupers' fields—easy picking for the resurrectionists. The wealthy, in contrast, were buried in sturdy triple caskets. Burial options multiplied as anxiety about body snatching spread: the so-called Patent Coffin of 1818 was explicitly and expensively marketed as being resurrectionist-proof. Cemeteries began offering a turn in the Dead House, where, for a fee, the closely guarded corpse would genteelly putrefy past the point of interest to dissectors, at which time it could be safely buried.

Meanwhile, the fears of the poor continued to mount, thanks in no small part to the appearance of a new verb in the lexicon: "burking," named for William Burke, the aging resurrectionist who pioneered the charming practice of luring beggars into his home for a charitable meal, then strangling them for a quick sell to the anatomists.* Although such indiscretions were generally overlooked by the august and compliant medical community, in the poor they induced a riotous displeasure. Frenzied crowds lynched captured resurrectionists, attacked the homes of anatomists, and burned hospitals.

---

*Ironic ending department: after Burke was caught and executed, his own skull was turned over to phrenologists for examination, in hopes of their detecting some scientific explanation for his heinous crime.

In New York City, shortly after the American Revolution, medical students from King's College (now Columbia University) were discovered digging for bodies in a paupers' graveyard (no doubt trying to cut the resurrectionists out as the middlemen and get some pocket money for themselves). The ensuing riot, known as the Doctor's Rebellion, lasted for days, forcing physicians to take shelter in the home of Alexander Hamilton while the state militia fired on the crowd.

In the eighteenth and nineteenth centuries many European governments took steps to control such mayhem and regulate the predation upon the poor. Anatomists were supplied with material, those unseemly burkers and grave robbers put out of business, and the poor kept in line, all with one handy new law: anyone who died destitute in a poorhouse or a paupers' hospital would be turned over to dissectors.

The historian Ruth Richardson, author of *Death, Dissection and the Destitute,* argues that the laws were designed to punish and terrorize the poor. Richardson points out that when the so-called Anatomy Act was first introduced in the British Parliament in 1829, it was soundly rejected as brutal and unfair. Yet just three years later, shortly after the Reform Riots instigated by the lower class, the act was reintroduced into the same Parliament and quickly passed by a vote of forty-six to four.

The Anatomy Act thus served to formalize the anatomists' overreliance on the corpses of the poor. In the century following the English enactment, for instance, more than 99 percent of the cadavers used in London for medical education came from poorhouses. Although the percentage has dropped in recent years, as more and more people voluntarily donate their bodies to research, it is still overwhelmingly the poor whose bodies provide the training base for physicians and scientists. I once spoke with an anthropologist who worked for one of the biological supply houses that sell skeletons and body parts to the medical community: His job was to comb rural India and buy skeletons from the poor relatives of the deceased. He never had trouble finding willing vendors.*

None of this should surprise. Socioeconomic inequity permeates every aspect of the health care system in the United States. A study conducted in early 1990 by Colin W. McCord and Harold P. Freeman, two surgeons at Harlem Hospital in New York, showed that African-American men in Harlem have a shorter life expectancy than men in Bangladesh. Other classic studies have demonstrated that in an ambulance trip to an emergency room, the poorer you appear to be, the more likely it is that you will be pronounced dead on arrival.

But poverty has a profound and even more direct influence on health. The ailments to which disadvantaged people typically succumb are quite unlike the ones that end the lives of the wealthy: chronic malnutrition, parasitic infections, lingering tuberculosis, wasting illnesses—in essence, diseases involving prolonged stress. Moreover, chronic stress induces numerous morphological changes in the body that can affect what is observed during autopsy.

Consider the adrenal gland. When the owner of the organ is in a fight-or-flight state, the adrenal gland secretes adrenaline and a class of steroids known as glucocorticoids, which speed up the heart rate and the metabolism. People who live in chronically stressful conditions place a constant demand on their adrenal glands to produce the hormones, and the glands may grow larger in compensation. Laboratory rats suffering from a chronic wasting disease display, at autopsy, adrenal glands larger than those of healthy control rats. Human cadavers show the same effect.

But medical scientists before the 1930s did not know this

---

*And there was often the disturbing suggestion that some individuals might even have hastened along the demise of their relatives for the sake of the transaction.

fact. Physicians, examining cadavers predominantly of the poor, thought they were learning what a normal adrenal gland looks like; instead, and unknowingly, they were observing the physiological effects of a lifetime of poverty. On the infrequent occasion when the body of someone with a higher income was examined, it was noted that the adrenal glands seemed oddly undersized—lighter in average weight than the ones mentioned in pathology journals and not at all like the ones observed in medical school.

Unfamiliar with the appearance of a normal adrenal gland, physicians invented a new disease to explain their discovery: idiopathic adrenal atrophy, which is a fancy way of saying that the adrenals had shrunk for some unknown reason. This "disease" flourished in the early twentieth century; then physicians caught on, and then, voilà, everyone was instantly cured of the malady.

The overreliance on anatomical material from the poor has led to other medical mishaps that are not so humorous. On at least one occasion, the distorted data base has had tragic consequences.

When a person is in a state of stress, the hormones secreted by the adrenal glands suppress the immune response. The chronic stress experienced by poverty-stricken people eventually can cause glands essential to the immune system, such as the thymus, located in the throat, to shrivel away to virtually nothing. As a result a disproportionately large amount of autopsy material—derived as it is from people with diseases caused by chronic stress and deprivation—included atrophied thymus glands. Thus what was regarded in the 1930s as a normal-sized thymus was actually a greatly shrunken one. The stage was set for a horrendous medical blunder.

For some time pediatricians had identified a relatively new disorder known colloquially as crib death and now called sudden infant death syndrome. Seemingly at random, parents who tucked their healthy infant into bed for a peaceful night's sleep would return in the morning to discover the infant had died. Modern medical scientists have made some progress in understanding SIDS: the disorder seems to occur in infants who in their third trimester as a fetus suffer an oxygen deficiency that damages the brain cells responsible for controlling respiration. At the time, however, physicians had no insight at all into the cause of the mysterious deaths.

Faced with this enigma, a pathologist named Paltauf, working at the end of the nineteenth century, adopted a logical course of research: He carefully autopsied SIDS infants and compared the results with autopsies of non-SIDS infants. Not surprisingly, Paltauf quickly spotted a whopper of a difference: The thymus glands of SIDS infants were far larger than those of the non-SIDS infants. It is perfectly obvious today what was occurring. The latter group, of course, had died of chronic, stressful illnesses that caused thymic atrophy, whereas the former had died suddenly. In examining the SIDS infants, Paltauf was the first pathologist to be systematically observing normal-sized thymuses.

But he had no way of knowing this, and Paltauf framed a hypothesis for the cause of SIDS that got normal and abnormal confused. In some infants, it seemed to him, the thymus was so abnormally enlarged that during sleep it pressed down on the trachea, suffocating the infant. By the turn of the century this disorder had a name, status thymicolymphaticus, and by the 1920s all the leading pediatric textbooks were offering the same advice: To prevent SIDS, infants' throats should be irradiated to shrink the menacing thymus. The treatment became a pediatric fad, persisting well into the 1950s. Irradiation of the thymus, of course, had no effect on the rate of SIDS. But next to the thymus, and equally exposed to radiation, was the thyroid gland, which helps control growth and metabolism. The spurious cure for this spurious disease eventually led to tens of thousands of cases of thyroid cancer.

It is a chilling experience to wander the dusty lower floor of a medical library, reading forgotten seventy-year-old pediatric texts with their dry discussions of status thymicolymphaticus. The technical details of the disorder, the plausible etiology, the photographs of the "enlarged" thymuses, the confident recommendation for treatment—all wrong, page after page.* What mistakes are we making now, in our modern scientific ignorance, and how many people will ultimately pay for it?

I won't climb up on my soapbox and preach about a world in which infants get small thymuses because they were born poor. Instead, I'll aim for something a bit more manageable. For starters, this piece of medical history teaches us something about the science we should be carrying out. We currently expend a great deal of effort on some amazing facets of medical research—sequencing the human genome, transplanting neurons, building artificial organs. This is great, but we still need smart people to study some moronically simple questions, like "How big is a normal thymus?" Because those questions often aren't all that simple. Maybe another lesson is that big confounds can come from subtle small places, a dictum understood by the best of public health researchers. It strikes me, however, that the most important lesson is one that transcends science and is relevant to our society at large, a society that can be immensely judgmental: Be *really* certain before you ever pronounce something to be the norm, because at that instant, you have now made it supremely difficult to ever again look at an exception to that supposed norm and to see it objectively.

---

*Amid this generally grim story, I was amused to note that by the late 1920s, this "disease" was so well established that new ground was being broken in describing the distinctive and striking behavioral features of infants who would later be found to have died of status thymicolymphaticus. They were characterized as having "phlegmatic" dispositions—presumably because these were normal kids and thus were phlegmatic about their imaginary illness.

FURTHER READING

The original observation of "enlarged" thymuses in SIDS infants was reported by A. Paltauf, "Plotzlicher Thymus Tod, Wiener klin," *Woeschesucher* (Berlin), nos. 46 and 9. The supposed disease was named a few years later in T. Escherich, "Status thymico-lymphaticus, Berlin klin," *Woeschesucher,* no. 29. An example of the typical 1920s textbook giving advice on preventing the disease with throat irradiation would be W. Lucas, *Modern Practise of Pediatrics* (New York: Macmillan, 1927); this is the text that contained the information on the personality profiles of infants with status thymicolymphaticus.

Lost amid this consensus of the savants was a 1927 study by E. Boyd ("Growth of the Thymus, Its Relation to Status Thymolymphaticus and Thymic Symptoms," *American Journal of Diseases of Children* 33: 867), which should have put the whole thing to rest. Boyd showed for the first time that a stressor (malnutrition, in this case) caused thymic shrinking. She demonstrated, moreover, that children who died in accidents turned out, upon autopsy, to be "suffering" from status thymicolymphaticus, and suggested for the first time that the whole thing might be an artifact. By the mid-1930s, the first studies had been conducted showing that an array of physical or psychological stressors would shrink the thymus, but it was not until 1945 that a leading textbook in the field emphatically stated that the disease was an artifact and that its treatment was a disaster: W. Nelson, *Nelson's Textbook of Pediatrics,* 4th ed. (Philadelphia: Saunders, 1945). And despite this information, the practice continued widely well into the 1950s.

Ruth Richardson's work can be found in her book *Death, Dissection and the Destitute* (London: Routledge and Kegan Paul, 1987). The relevance of socioeconomic status to the likelihood of your being revived during an ambulance ride is discussed in D. Sudnow, *Passing On: The Social Organization of Dying* (Englewood Cliffs, N.J.: Prentice-Hall, 1967). The finding of a shorter life expectancy among African-American men in Harlem than among men in Bangladesh is reported in C. McCord, "Poverty, Race, Racism, and Survival," *Annals of Epidemiology* 3 (1993): 145.

Junk Food Monkeys

David Gilhooly, Tall Excessive Sundae, *1991;*
courtesy John Natsoulas Press, Davis, California

When's the last time you thought about Jean-Jacques Rousseau? Remember his noble savage, his idealized view of mankind in its primordial splendor, when life was gentle, innocent, and natural? Not that most of us reject Rousseau's thinking; it's just that amid the bustle and ambition of the remnants of the Me Generation, it no longer seems fashionable to ponder the possibly superior moral state of primordial humans.

As for their possibly superior physical state—now that's an issue of more than passing concern, and here a Rousseauian view of sorts continues to hold sway. As a society that spends billions of dollars on medical care, health clubs, and looking good, what we want to know is: How was the physical health of primordial humans? What was their secret workout regimen for achieving their beautiful precivilized bodies? What were cholesterol levels like in the Garden of Eden?

Many think that in certain respects early humans fared better than we do. When you subtract the accidents, infections, and infantile diseases that beset them, our forebears might not have had it so bad. In a 1985 article in the *New England Journal of Medicine,* physician-anthropologists S. Boyd Eaton and Melvin Konner did an ingenious job of reconstructing the likely diet of our Paleolithic ancestors, and they concluded that there was much to be said for the high-fiber, low-salt, low-fat diet.

Anthropologists studying the hunter-gatherers in the Kalahari Desert, a group that's believed to retain many of the ways of life of earlier humans, have come to similar conclusions. Among the !Kung tribe, for example, some maladies that we view as a normal part of human aging are proving not to be obligatory after all: hearing acuity is not lost; blood pressure and cholesterol levels do not rise; degenerative heart disease doesn't seem to develop. As we sit here amid our ulcers and hypertension and hardening arteries, it is getting harder for us to avoid the uncomfortable suspicion that we have fallen from a state of metabolic grace.

For the past decade I have been observing a group of some of our closest relatives as they fell from their own primordial metabolic grace into something resembling our nutritional decadence. My subjects are the olive baboons that I study during my summers in the Masai Mara National Reserve in the Serengeti Plain of Kenya. Chiefly I'm interested in the relationships between their social behavior and dominance rank, the amount of social stress the baboons experience, and how their bodies react to it. To study the questions I am interested in, I have to combine extensive behavioral observations with some basic lab work: drawing the animals' blood, measuring hormones, monitoring blood pressure, and conducting other clinical tests to find out how their bodies are functioning. And in this context Rousseau reared his worrisome head.

Masai Mara is a wonderful place to be a researcher. And it's a fairly idyllic place to be a baboon, a vast untouched landscape of savannahs and woodlands and one of the last great refuges for wild animals left on earth. Herds of wildebeest roam on the open plains, lions lounge beneath the flat-topped acacia trees, giraffe and zebra drink side by side at the watering places. Inevitably Masai Mara has also become an attractive place to tourists, resulting in all the usual problems that occur when large influxes of people descend on previously virgin wilderness.

One of the biggest problems here, as in our own national parks, is what to do with the garbage. The solution so far has been to dump much of it into large pits, five feet deep by thirty feet wide, hidden among trees in out-of-the-way areas. Brimming with food and refuse rotting in ninety-degree heat, infested with flies and circled by vultures and hyenas, the pits look like a scene from Hieronymus Bosch. And about a decade ago, one of the baboon troops that I study had such a Garden of Earthly Delights dumped right in the middle of its territory.

For the baboons this was a major change in fortune, the primatological equivalent, perhaps, of winning the lottery. A major concern in any wild animal's life is obtaining food, and an average baboon in the Serengeti spends 30 to 40 percent of each day foraging—climbing trees to reach fruits and leaves, digging laboriously in the ground to unearth tubers, walking five or ten miles to reach sources of food. Their diet is spartan: figs and olives, grass and sedge parts, corms, tubers, and seedpods. It's unusual for them to hunt or scavenge, and meat accounts for less than 1 percent of the food they consume. So the typical baboon diet teems with fiber and is very low in fat, sugar, and cholesterol.

For the nouveau riche Garbage Dump troop, life changed dramatically. When I started observing them, in 1978, they had recently discovered the dump's existence and were making an occasional food run. By 1980 the entire group—some eighty animals, ranging from twenty-five-year-old adults to newborn infants—had moved into new sleeping quarters in the trees surrounding the dump. Instead of stirring at dawn, these animals would typically stay in the trees, snoozing and grooming, and only rouse themselves in time to meet the

9 A.M. garbage tractor. The day's feeding would be finished after half an hour of communal frenzy over the pickings. But it was the pickings themselves that made the biggest difference in the baboons' lives.

Once, in the name of science, I donned lab gloves, held my breath, and astonished the tractor driver by methodically sifting through his moldering garbage. The refuse was certainly a far cry from tubers and leaves: fried drumsticks or a slab of beef left over by a tourist with eyes bigger than his stomach; fruit salad gone a bit bad, perhaps left too long on the sun-drenched buffet table; fragments of pies and cakes, and alarming yellow dollops of custard pudding, nibbled at by a disciplined dieter—processed sugars, fat, red meat, and cholesterol, our modern Four Horsemen of the Apocalypse.

And what were the physiological consequences for these baboons in Utopia? First the good news: Young baboons grew faster, reaching developmental landmarks such as puberty at earlier ages. These beneficial changes were exactly what one would expect of humans switching from a lean subsistence diet to a more affluent Westernized fare. In the countries of the West the age of first menstruation has declined from an estimated fifteen years during the 1800s to our current average of about twelve and a half, and diet is thought to have much to do with that.

The trend in baboons has been particularly well documented by Jeanne Altmann of the University of Chicago, a biologist studying both foraging and garbage-eating troops in another park in Kenya. Among her animals, eating garbage has led to the onset of puberty at age three and a half instead of age five. Females now typically give birth for the first time at age five, a year and a half earlier than before. Moreover, because the infants develop faster, they are weaned earlier; consequently, females start menstruating again that much sooner, and once they resume, they conceive more quickly.

Indeed, Altmann's garbage eaters have had something of a baby boom compared with their foraging cousins.

Another advantage of garbage eating became clear during the tragic East African drought of 1984. During that period wild game found life extremely difficult. The luckier animals merely spent more time and covered more distance in search of food. The less lucky starved and succumbed to diseases previously held in check. However, tourists did not starve, and neither did baboons living off their detritus.

So, at first glance, from the evolutionary standpoint of reproductive fitness, some daily custard pudding appears to do wonders, increasing reproductive rates and buffering the troop from famine. But now here's the bad news: some of the same lousy changes seen in humans eating Westernized fare also occurred in baboons.

Your average wild baboon eating a natural diet has cholesterol levels that would shame the most ectomorphic triathlete. University of Texas pathologist Glen Mott and I have studied a number of troops and found cholesterol levels averaging an amazing 66 milligrams in 100 cubic centimeters of blood among adult males. Not only that, but more than half the total cholesterol was in the form of high-density lipoproteins, the "good" type. In humans, cholesterol levels less than 150 with a third of the total in the high-density form are grounds for bragging at the health club.

But when we studied the Garbage Dump baboons, a different picture emerged. Cholesterol levels were nearly a third higher, and most of the increase was attributable to a rise in damaging low-density lipoproteins, the type that builds up plaque on artery walls. Joseph Kemnitz, a primatologist at the Wisconsin Regional Primate Research Center, analyzed blood samples from these animals and found that levels of insulin were more than twice as high in the Garbage Dumpers as in those eating a natural diet. This hormone is secreted by the pancreas in response to eating, especially eating rich, sug-ary food, and its function is to tell cells to store glucose for future use as energy.

If insulin levels rise too high, however, cells become inured to its message; instead of being stored, glucose is left circulat-ing in the bloodstream. It's this state of affairs that can even-tually lead to adult-onset diabetes, a distinctly Western malady. Since the Garbage Dumpers came from gene stocks similar to those of the natural foragers, genetic differences couldn't account for their much higher insulin levels. The most likely suspect was their junk food diet and their relative inactivity.

Are the Garbage Dumpers now at risk for diabetes and heart disease, candidates for celebrity diets and coronary bypasses? It is hard to tell; no one has ever reported adult-onset diabetes in wild baboons, but, then, no one has ever looked. Surprisingly, people have looked at the cardiovascular systems of wild baboons and found fatty deposits in their blood vessels and hearts. Garbage Dumpers presumably run a higher risk of depositing fat in their arteries, but whether it will affect their health and life spans remains to be seen. One major impact on the health of Garbage Dumpers became clear in a grim way, however. If you are going to spend your time around human garbage, you'll have to deal with what-ever that garbage is infected with. And if these infectious agents are new to you, your immunological defenses are not likely to be very good.

A few years ago some of my Garbage Dump animals became dramatically ill. They wasted away, coughing up blood, losing the use of their limbs. Three veterinarians from the Institute of Primate Research in Nairobi—Ross Tarara, Mbaruk Suleman, and James Else—joined me to investigate the outbreak and traced it to bovine tuberculosis, probably from eating contaminated meat. It was the first time the dis-ease had been reported in wild primates, and by the time it abated it had killed half the Garbage Dump troop. For them, there had been no free lunch after all.

Has the Westernization of these animals' diets thus been good or bad? This is much the same as questioning the benefits of our own Westernization. Toxic wastes and automatic weapons strike me as bad developments; on the other hand, vaccines, thermal underwear, and eighty-year life expectancies seem like marvelous improvements upon the Middle Ages. All things considered, we seem to have benefited, at least from a health perspective.

For the baboons, too, the answer must be carefully considered. Growing fat is unwise if you plan to sit in an arboreal armchair, but very wise if you're about to face a dry season. Similarly, a nice piece of meat is a fine thing during a famine, but not such a hot idea if it's contaminated.

It is platitudinous to say that this is a complicated issue, but it is clear that life for the foraging baboon is not one of pure Rousseauian ease, nor life for the garbage eaters one of unambiguous decline. My bias is that the latter's health has, on the whole, suffered from their garbage trove. But what has struck me in these studies is that there are also benefits and the judgment is somewhat difficult to make. It seems that even for a baboon, just as for us, there are few unambiguous rules for figuring out what to do with the choices life throws in our laps.

FURTHER READING

The salutary diet of our ancestors is discussed in S. Eaton and M. Konner, "Paleolithic Nutrition: A Consideration of Its Nature and Current Implications," *New England Journal of Medicine* 312 (1985): 283. The reader might also want to check a more accessible version of the same, namely: S. Eaton, M. Shostak, and M. Konner, *The Paleolithic Prescription: A Program of Diet and Exercise and a Design for Living* (New York: Harper and Row, 1988).

Cholesterol levels in wild baboons is considered in R. Sapolsky and G. Mott, "Social Subordinance in Wild Baboons Is Associated with Suppressed High-Density Lipoprotein-Cholesterol Concentrations: The Possible Role of Chronic Social Stress," *Endocrinology* 121 (1987): 1605.

The effects of diet on metabolic profiles in wild baboons is covered in J. Altmann, D. Schoeller, S. Altmann, P. Muruthi, and R. Sapolsky, "Variability in Body Size and Fatness in a Free-Living Nonhuman Primate Population," *American Journal of Primatology* 30 (1993): 149.

The vulnerability of primates living on garbage diets to infectious diseases is covered in depressing detail in two papers: R. Tarara, M. Suleman, R. Sapolsky, M. Wabomba, and J. Else, "Tuberculosis in Wild Baboons *(Papio cynocephalus)* in Kenya," *Journal of Wildlife Disease* 21 (1985): 137; and in R. Sapolsky and J. Else, "Bovine Tuberculosis in a Wild Baboon Population: Epidemiological Aspects," *Journal of Medical Primatology* 16 (1987): 229.

The Burden
of Being Burden-Free

Clifford Goodenough, Figure Walking in a Landscape, *1991;*
courtesy Davidson Galleries, Seattle

After grad school, I spent a few years doing research at the Salk Institute in San Diego, a famously magnificent structure perched on the edge of a cliff overlooking the Pacific. Next door was a hang gliding center, and its denizens would often glide past the windows of my upper-floor lab. For some reason, I had decided that the sport was idiotic, its practitioners a bunch of irritating cowboys. One morning, I glanced up from, perhaps, my thousandth pipetting of the day, and my eyes briefly locked with those of a passing hang glider. It occurred to me in that instant that we were having the identical thought: "No way in hell would I like to be in your shoes, buddy."

One person's pleasure, another's poison. This fool had paid good money to glide around the cliffs, while I'd do the same to get to spend sixteen hours a day toiling away at my science. And for each of us, the other's activity would probably count as a stressful misery.

It is quite clear that you have a major medical problem on your hands if you are gored by an elephant or if you contract dengue fever. But it is far from clear if, instead, you are subjected to years of traffic jams, mortgages, deadlines, or relationship problems—maybe they represent profound psychological burdens, or maybe they are things you barely notice. We differ tremendously as to what psychologically stresses us. Obviously, we also differ tremendously as to the diseases we get and the courses they take. And there are reasons to think that how much we are buffeted with stressors may have something to do with what sorts of diseases afflict us.

A lot of research in health psychology has explored the immensely important fact that certain personality types seem to be predisposed toward stress-related disease, or toward secreting too much of certain hormones that can cause stress-related disease. The most frequently studied of these hormones are adrenaline and glucocorticoids. These hormones are secreted directly into the bloodstream and help you adapt to an acute stressful crisis by mobilizing the fight-or-flight response. However, if these same hormones are secreted chronically because of psychological stress, as seen in certain personality types, they can cause a variety of maladies. And some individuals with these personality types already show the stigmata of those maladies—elevated blood pressure, irritable bowels, reproductive suppression, irregular sleep. Why are their bodies working differently from most people's?

Most broadly, a stressor can be defined as a physical or psychological challenge that upsets the body's equilibrium,* and coping responses are what help reestablish such equilibrium. Obviously, one would like there to be an appropriate match between the *type* of stressor and the *type* of response. For example, it is generally a wise thing to come up with a response such as "I know what I did wrong, and I can avoid this happening again" when confronted with a bad test score, and to come up with a massive diversion of energy to your thigh muscles when confronted with a predatory lion—and not the other way around. In addition, one wants a fairly good match between the intensity or duration of stressors and responses—

---

*A more jargony term for that equilibrium, one that everyone got tossed at them in ninth-grade biology, would be "homeostasis."

a mild threat should not involve vigilant hypertension for hours afterward. Framed these ways, a common theme among many of these personality types that are predisposed toward stress-related diseases is a discrepancy between the stressor a person is pummeled with and the style of coping response mobilized.

One example of such maladaptiveness would be people with a major depression. A classic picture of many depressives is that, as a result of life's painful lessons, they have learned to be helpless. When faced with stressful challenges, they give up before things have even started. They don't attempt to cope or, if they happen to stumble onto something effective, don't recognize it for what it is. And, as has been known for decades, such depressives often have elevated stress hormone levels in their bloodstream.

A similar physiologic profile is seen in people who are at the opposite end of the spectrum when it comes to coping with things the wrong way. People with anxiety disorders can be thought of as persistently mobilizing coping responses that are disproportionately large. For them, life is filled with threats around every corner, threats that demand a constant hyper-vigilance, an endless skittering search for safety, a sense that the rules are constantly changing. A similar picture occurs in "Type A" individuals, but with a twist. Type A represents perhaps the most durable and studied of personality types that carry a disease risk. As initially characterized in the 1960s by the cardiologists Meyer Friedman and Ray Rosenman, Type A individuals are immensely competitive, overachieving, time-pressured, impatient, and hostile. A zillion studies later, it has become clear that the key component of Type A personality is the hostility—they perceive a need for coping responses that are disproportionately large, and their coping responses are disproportionately likely to be angry ones. Give a Type A individual an unsolvable puzzle, make them role-play an interpersonal conflict, expose them without their knowledge to a research confederate who (unbeknownst to them) deliberately botches a job . . . and the Type A individual boils. They view every frustration as an intentional, personal, malevolent affront . . . and their bloodstream fills with stress hormones and their blood pressure soars. For them, a lifetime of such corrosive discrepancy between stressor and response can eventually exact a major cardiovascular price.

Hypertension also stalks a lesser-known personality type, one driven by what Sherman A. James, an epidemiologist at the University of Michigan, calls John Henryism. The name refers to the American folk hero who, hammering a six-foot-long steel drill, tried to outrace a steam drill tunneling through a mountain. John Henry beat the machine, only to fall dead from the superhuman effort. As James defines it, John Henryism involves the belief that any and all demands can be vanquished, so long as you work hard enough. On questionnaires, John Henry individuals strongly agree with statements such as "When things don't go the way I want them, it just makes me work even harder," or "Once I make up my mind to do something, I stay with it until the job is completely done." This is the epitome of individuals with, to use some psychological jargon, an internal locus of control—they believe that, with enough effort and determination, they can regulate all outcomes.

What's so wrong with that? Nothing, if you have the good fortune to live in the privileged, meritocratic world in which one's efforts truly do have something to do with the rewards one gets, and in a comfortable, middle-class world, an internal locus of control does wonders. However, in a world of people born into poverty with limited educational or occupational opportunities facing prejudice and racism, it can be a disaster to be a John Henry, to decide that those insurmountable odds could have been surmounted, if only you could have worked *even* harder. Stress physiologists have long known the dangers of trying to control the uncontrollable, a classic case of mismatched coping. Strikingly, James's pioneering work has shown the cardiovascular disease dangers of John Henryism to occur predominately among the very people who most resemble the mythic John Henry himself, working-class African-Americans.

These links between personality type and overactive stress responses have been known for years. Yet some recent studies reveal an additional, unlikely personality type that has this sort of physiologic problem.

These are not people who are dealing with their stressors too passively, too persistently, too vigilantly, or with too much hostility. They don't appear to have all that many stressors. They claim they're not depressed or anxious, and the psychological tests they are given show they're right. In fact, they describe themselves as pretty happy, successful, and accomplished (and, according to personality tests conducted by psychologist Dirk H. Hellhammer and colleagues at Trier University in Germany, they really are all the above). Yet these people (comprising approximately 5 percent of the population) have chronically activated stress responses. What's their problem?

Their problem, I think, is one that offers insight into an unexpected vulnerability of our human psyche. The individuals in question are said to have "repressive" personalities, and we all have met someone like them. In fact, we usually regard these folks with a twinge of envy—"I wish I had their discipline; everything seems to come so easily to them. How do they do it?"

These are the archetypal people who cross all their T's and dot all their I's. They describe themselves as planners who don't like surprises, who live structured, rule-bound lives—walking to work the same way each day, always wearing the same style of clothes; the sort of people who can tell you what they're having for lunch two weeks from Wednesday. Not surprisingly, they don't like ambiguity and strive to set up their world in black and white, filled with good or bad people, behaviors that are permitted or strictly forbidden. And they keep a tight lid on their emotions. Stoic, regimented, hardworking, productive, solid folks who never stand out in a crowd (unless you begin to wonder at the unconventional nature of their extreme conventionality).

Some personality tests, pioneered by the psychologist Richard J. Davidson and colleagues at the University of Wisconsin, identify repressive individuals. For starters, as noted, the personality tests show that these people aren't depressed or anxious. Instead, the tests reveal their need for social conformity, their dread of social disapproval, and their discomfort with ambiguity, as shown by the extremely high rates at which they agree with statements framed as absolutes, statements filled with the words "never" and "always." No gray tones here.

Intertwined with those characteristics is a peculiar lack of emotional expression. The tests reveal how repressive people "inhibit negative affect"—no expressing of those messy, complicated emotions for them. This was shown with particular subtlety in a study where repressive and nonrepressive individuals were asked to recall an experience associated with a specific, strong emotion. Both groups reported that particular emotion with equal intensity. However, when asked what else they were feeling, nonrepressors would typically report an array of additional, nondominant feelings: "Well, it mostly made me angry, but also a little sad, and a little disgusted too. . . ." Repressors steadfastly reported no secondary emotions. Black and white feelings, with little tolerance for subtle blends. As another example of that, repressors are less likely to report observing a blend of dominant and nondominant emotions in the facial expressions of other people.

Emotions kept firmly under control, "never" and "always" to sort out those ambiguous curveballs that life throws at you now and then, and things keep flowing without a ripple.

Are these people for real? Maybe not. Maybe beneath their tranquil exteriors, they're actually anxious messes who won't admit to their frailties. Careful study indicates that some repressors are indeed mostly concerned about keeping up appearances. (One clue is that they tend to give less "repressed" answers on personality questionnaires when they can be anonymous.) And so their physiologic symptoms of stress are easy to explain: cross those folks off the list.

What about the rest? Could they be deceiving themselves—roiling with anxiety, but not even aware of it? Even careful questionnaires cannot detect that sort of self-deception; to ferret it out, psychologists traditionally rely on less structured, more open-ended tests (of the "What do you see in this picture?" variety). Those tests show that, yes, some repressors are far more anxious than they realize; their physiological stress is also readily explained.

Yet even after you cross the anxious self-deceivers off the list, there remains a group of people with tight, constrained personalities who are truly just fine; mentally healthy, happy, productive, socially interactive. But they have overactive stress responses. A 1996 paper by psychologists Laurel L. Brown, Andrew J. Tomarken, and colleagues at Vanderbilt University shows that repressives have levels of glucocorticoids in their bloodstream that are as elevated as those of highly anxious individuals. Earlier work by Davidson showed that the fight-or-flight component of the nervous system (the source of adrenaline) is also overactive. When exposed to a cognitive challenge, repressors show unusually large increases in heart rate, blood pressure, sweating, and muscle tension. And these overaroused stress responses may exact a price. For example, coronary disease patients who have repressive personalities are more vulnerable to cardiac complications than are nonrepressors.

Overactive, endangering stress responses—yet the people harboring them are not stressed, depressed, or anxious. Back to our envious thought—"I wish I had their discipline. How do they do it?" The way they do it, I suspect, is by working like maniacs to generate their structured, repressed worlds with no ambiguity or surprises. And this comes with a physiological bill.

Nothing brings home the material nature of our minds like a certain type of study utilizing a PET scanner, an instrument that measures the consumption of glucose or oxygen in different parts of the brain. Give a person a mathematical task, and the PET scan indicates that metabolic demand increases in certain areas of the brain. Have them recall a happy memory, and a different anatomic pattern lights up. Provoke anxiety, and there's a different pattern. *It takes energy to think.* And, apparently, it takes a whole lot of energy to think and plan and avoid and deny all the time like a repressor. Tomarken and Davidson have used electroencephalographic (EEG) techniques to show unusually enhanced activity in a portion of the frontal cortex of repressors. This is a region of the brain involved in inhibiting impulsive emotion and cognition, the nearest neuroanatomical equivalent we have to a superego.

A similar theme emerges from the work of James J. Gross at Stanford University. Show people film clips that evoke strong emotions (such as graphic footage of an amputation) and you trigger some pretty strong arousal of the fight-or-flight response. Show people the same film but now instruct them to try to hide their emotions, and the neural response is even stronger, just like for a repressor. It begins to seem plausible that there is a physiological cost to going through life with all these psychic sphincter muscles holding on for dear life.

This is a newly emerging area of research, and much

remains unknown. For example, how readily can a repressive individual become less so with therapy? When they change, do their hormone profiles change as well? Will it ever be possible to discern a repressive individual just by recognizing a distinctive hormonal profile? Regardless of the eventual answers to these questions, the physiologic correlates of repressive personality teach us some pretty interesting and unexpected things right now.

It can be a frightening world out there, and the body may well reflect the effort of threading our way through those dark, menacing forests. How much better it would be to be able to sit, relaxed, on the sun-drenched porch of a villa, far, far away from the wild things baying. Yet what looks like relaxation could well be exhaustion—exhaustion from the labor of having built a wall around that villa, the effort of keeping out that unsettling, challenging, vibrant world. A lesson of repressive personality types and their invisible burdens is that, sometimes, it can be enormously stressful to construct a world without stressors.

FURTHER READING

The relationship between depression and an overactive stress response is reviewed in R. Sapolsky and P. Plotsky, "Hypercortisolism and Its Possible Neural Bases," *Biological Psychiatry* 27 (1990): 937. Depression as a disorder of inappropriate coping is reviewed in the classic M. Seligman, *Helplessness: On Development, Depression and Death* (New York: W. H. Freeman, 1975). The current thinking about Type A is reviewed in R. Williams, *The Trusting Heart: Great News about Type A Behavior* (New York: Random House, 1989). A review of John Henryism can be found in S. James, "John Henryism and the Health of African-Americans," *Culture, Medicine and Psychiatry* 18 (1994): 163.

The studies concerning repressive personality and its physiologic correlates can be found in:

J. Brandtstadter, B. Baltes-Gotz, C. Kirschbaum, and D. Hellhammer,

"Developmental and Personality Correlates of Adrenocortical Activity as Indexed by Salivary Cortisol: Observations in the Age Range of 35 to 65 Years," *Journal of Psychosomatic Research* 35 (1991): 173. This is the study of Hellhammer and colleagues.

L. Brown, A. Tomarken, D. Orth, P. Loosen, N. Kalin, and R. Davidson, "Individual Differences in Repressive-Defensiveness Predict Basal Salivary Cortisol Levels," *Journal of Personality and Social Psychology* 70 (1996): 362.

R. Shaw, F. Cohen, R. Fishman-Rosen, M. Murphy, S. Stertzer, D. Clark, and K. Myler, "Psychologic Predictors of Psychosocial and Medical Outcomes in Patients Undergoing Coronary Angioplasty," *Psychosomatic Medicine* 48 (1986): 582.

R. Shaw, F. Cohen, B. Doyle, and J. Palesky, "The Impact of Denial and Repressive Style on Information Gain and Rehabilitation Outcomes in Myocardial Infarction Patients," *Psychosomatic Medicine* 47 (1985): 262.

A. Tomarken and R. Davidson, "Frontal Brain Activation in Repressors and Nonrepressors," *Journal of Abnormal Psychology* 103 (1994): 339.

D. Weinberger, G. Schwartz, and R. Davidson, "Low-Anxious, High-Anxious, and Repressive Coping Styles: Psychometric Patterns and Behavioral and Physiological Responses to Stress," *Journal of Abnormal Psychology* 88 (1979): 369.

The studies of James Gross concerning the intentional inhibition of emotional expression can be found in J. Gross and R. Levenson, "Emotional Suppression: Physiology, Self-Report, and Expressive Behavior," *Journal of Personality and Social Psychology* 64 (1993): 870. See also: J. Gross and R. Levenson, "Hiding Feelings: The Acute Effects of Inhibiting Negative and Positive Emotion," *Journal of Abnormal Psychology* (in press).

The Trouble with Testosterone

WILL BOYS JUST BE BOYS?

Max Ernst, Health Through Sports, *c. 1920;*
The Menil Collection, Houston

Face it, we all do it. We all believe in certain stereotypes about certain minorities. The stereotypes are typically pejorative and usually false. But every now and then, they are true. I write apologetically as a member of a minority about which the stereotypes are indeed true. I am male. We males account for less than 50 percent of the population, yet we generate an incredibly disproportionate percentage of the violence. Whether it is something as primal as having an ax fight in an Amazonian clearing or as detached as using computer-guided aircraft to strafe a village, something as condemned as assaulting a cripple or as glorified as killing someone wearing the wrong uniform, if it is violent, males excel at it.

Why should that be? We all think we know the answer. A dozen millennia ago or so, an adventurous soul managed to lop off a surly bull's testicles and thus invented behavioral endocrinology. It is unclear from the historical records whether this individual received either a grant or tenure as a result of this experiment, but it certainly generated an influential finding—something or other comes out of the testes that helps to make males such aggressive pains in the ass.

That something or other is testosterone.* The hormone

---

*Testosterone is one of a family of related hormones, collectively known as "androgens" or "anabolic steroids." They all are secreted from the testes binds to specialized receptors in muscles and causes those cells to enlarge. It binds to similar receptors in laryngeal cells and gives rise to operatic basses. It causes other secondary sexual characteristics, makes for relatively unhealthy blood vessels, alters biochemical events in the liver too dizzying to even contemplate, has a profound impact, no doubt, on the workings of cells in big toes. And it seeps into the brain, where it binds to those same "androgen" receptors and influences behavior in a way highly relevant to understanding aggression.

What evidence links testosterone with aggression? Some pretty obvious stuff. Males tend to have higher testosterone levels in their circulation than do females (one wild exception to that will be discussed later) and to be more aggressive. Times of life when males are swimming in testosterone (for example, after reaching puberty) correspond to when aggression peaks. Among numerous species, testes are mothballed most of the year, kicking into action and pouring out testosterone only during a very circumscribed mating season—precisely the time when male-male aggression soars.

Impressive, but these are only correlative data, testosterone repeatedly being on the scene with no alibi when some aggression has occurred. The proof comes with the knife, the performance of what is euphemistically known as a "subtraction" experiment. Remove the source of testosterone in species after species and levels of aggression typically plummet. Reinstate normal testosterone levels afterward with injections of synthetic testosterone, and aggression returns.

To an endocrinologist, the subtraction and replacement

---

or are the result of a modification of testosterone, they all have a similar chemical structure, and they all do roughly similar things. Nonetheless, androgen mavens spend entire careers studying the important differences in the actions of different androgens. I am going to throw that subtlety to the wind and, for the sake of simplification that will horrify many, will refer throughout to all of these related hormones as "testosterone."

paradigm represents pretty damning proof: this hormone is involved. "Normal testosterone levels appear to be a prerequisite for normative levels of aggressive behavior" is the sort of catchy, hummable phrase that the textbooks would use. That probably explains why you shouldn't mess with a bull moose during rutting season. But that's not why a lot of people want to understand this sliver of science. Does the action of this hormone tell us anything about *individual* differences in levels of aggression, anything about why some males, some human males, are exceptionally violent? Among an array of males—human or otherwise—are the highest testosterone levels found in the most aggressive individuals?

Generate some extreme differences and that is precisely what you see. Castrate some of the well-paid study subjects, inject others with enough testosterone to quadruple the normal human levels, and the high-testosterone males are overwhelmingly likely to be the more aggressive ones. However, that doesn't tell us much about the real world. Now do something more subtle by studying the normative variability in testosterone—in other words, don't manipulate anything, just see what everyone's natural levels are like—and high levels of testosterone and high levels of aggression still tend to go together. This would seem to seal the case—interindividual differences in levels of aggression among normal individuals are probably driven by differences in levels of testosterone. But this turns out to be wrong.

Okay, suppose you note a correlation between levels of aggression and levels of testosterone among these normal males. This could be because *(a)* testosterone elevates aggression; *(b)* aggression elevates testosterone secretion; *(c)* neither causes the other. There's a huge bias to assume option *a,* while *b* is the answer. Study after study has shown that when you examine testosterone levels when males are first placed together in the social group, testosterone levels predict nothing about who is going to be aggressive. The subsequent behavioral differences drive the hormonal changes, rather than the other way around.

Because of a strong bias among certain scientists, it has taken forever to convince them of this point. Behavioral endocrinologists study what behavior and hormones have to do with each other. How do you study behavior? You get yourself a notebook and a stopwatch and a pair of binoculars. How do you measure the hormones? You need a gazillion-dollar machine, you muck around with radiation and chemicals, wear a lab coat, maybe even goggles—the whole nine yards. Which toys would you rather get for Christmas? Which facet of science are you going to believe in more? Because the endocrine aspects of the business are more high-tech, more reductive, there is the bias to think that it is somehow more scientific, more powerful. This is a classic case of what is often called physics envy, the disease among scientists where the behavioral biologists fear their discipline lacks the rigor of physiology, the physiologists wish for the techniques of the bio-chemists, the biochemists covet the clarity of the answers revealed by the molecular biologists, all the way down until you get to the physicists, who confer only with God.* Hor-mones seem to many to be more real, more substantive, than the ephemera of behavior, so when a correlation occurs, it

---

*An example of physics envy in action. Recently, a zoologist friend had obtained blood samples from the carnivores that he studies and wanted some hormones in the sample assays in my lab. Although inexpe-rienced with the technique, he offered to help in any way possible. I felt hesitant asking him to do anything tedious but, so long as he had offered, tentatively said, "Well, if you don't mind some unspeakable drudgery, you could number about a thousand assay vials." And this scientist, whose superb work has graced the most prestigious science journals in the world, cheerfully answered, "That's okay, how often do I get to do *real* science, working with test tubes?"

must be because hormones regulate behavior, not the other way around.

As I said, it takes a lot of work to cure people of that physics envy, and to see that interindividual differences in testosterone levels don't predict subsequent differences in aggressive behavior among individuals. Similarly, fluctuations in testosterone levels within one individual over time do not predict subsequent changes in the levels of aggression in that one individual—get a hiccup in testosterone secretion one afternoon and that's not when the guy goes postal.

Look at our confusing state: normal levels of testosterone are a prerequisite for normal levels of aggression, yet changing the amount of testosterone in someone's bloodstream within the normal range doesn't alter his subsequent levels of aggressive behavior. This is where, like clockwork, the students suddenly start coming to office hours in a panic, asking whether they missed something in their lecture notes.

Yes, it's going to be on the final, and it's one of the more subtle points in endocrinology—what is referred to as a hormone having a "permissive effect." Remove someone's testes and, as noted, the frequency of aggressive behavior is likely to plummet. Reinstate precastration levels of testosterone by injecting that hormone, and precastration levels of aggression typically return. Fair enough. Now this time, castrate an individual and restore testosterone levels to only 20 percent of normal and . . . amazingly, normal precastration levels of aggression come back. Castrate and now generate twice the testosterone levels from before castration—and the same level of aggressive behavior returns. You need some testosterone around for normal aggressive behavior—zero levels after castration, and down it usually goes; quadruple it (the sort of range generated in weight lifters abusing anabolic steroids), and aggression typically increases. But anywhere from roughly 20 percent of normal to twice normal and it's all the same; the brain can't distinguish among this wide range of basically normal values.

We seem to have figured out a couple of things by now. First, knowing the differences in the levels of testosterone in the circulation of a bunch of males will not help you much in figuring out who is going to be aggressive. Second, the subtraction and reinstatement data seem to indicate that, nevertheless, in a broad sort of way, testosterone causes aggressive behavior. But that turns out not to be true either, and the implications of this are lost on most people the first thirty times you tell them about it. Which is why you'd better tell them about it thirty-one times, because it is the most important point of this piece.

Round up some male monkeys. Put them in a group together, and give them plenty of time to sort out where they stand with each other—affiliative friendships, grudges and dislikes. Give them enough time to form a dominance hierarchy, a linear ranking system of numbers 1 through 5. This is the hierarchical sort of system where number 3, for example, can pass his day throwing around his weight with numbers 4 and 5, ripping off their monkey chow, forcing them to relinquish the best spots to sit in, but, at the same time, remembering to deal with numbers 1 and 2 with shit-eating obsequiousness.

Hierarchy in place, it's time to do your experiment. Take that third-ranking monkey and give him some testosterone. None of this within-the-normal-range stuff. Inject a ton of it into him, way higher than what you normally see in a rhesus monkey; give him enough testosterone to grow antlers and a beard on every neuron in his brain. And, no surprise, when you then check the behavioral data, it turns out that he will probably be participating in more aggressive interactions than before.

So even though small fluctuations in the levels of the hormone don't seem to matter much, testosterone still causes aggression. But that would be wrong. Check out number 3 more closely. Is he now raining aggressive terror on any and all in the group, frothing in an androgenic glaze of indiscriminate violence? Not at all. He's still judiciously kowtowing to numbers 1 and 2, but has simply become a total bastard to numbers 4 and 5. This is critical: testosterone isn't *causing* aggression, it's *exaggerating* the aggression that's already there.

Another example just to show we're serious. There's a part of your brain that probably has lots to do with aggression, a region called the amygdala.* Sitting right near it is the Grand Central Station of emotion-related activity in your brain, the hypothalamus. The amygdala communicates with the hypothalamus by way of a cable of neuronal connections called the stria terminalis. No more jargon, I promise. The amygdala has its influence on aggression via that pathway, with bursts of electrical excitation called action potentials that ripple down the stria terminalis, putting the hypothalamus in a pissy mood.

Once again, do your hormonal intervention; flood the area with testosterone. You can do that by injecting the hormone into the bloodstream, where it eventually makes its way to this part of the brain. Or you can be elegant and surgically microinject the stuff directly into this brain region. Six of one, half a dozen of the other. The key thing is what doesn't happen next. Does testosterone now cause there to be action potentials surging down the stria terminalis? Does it turn on that pathway? Not at all. If and only if the amygdala is *already* sending an aggression-provoking volley of action potentials down the stria terminalis, testosterone increases the rate of such action potentials by shortening the resting time between

---

*And no one has shown that differences in the size or shape of the amygdala, or differences in the numbers of neurons in it, can begin to predict differences in normal levels of aggression. Same punch line as with testosterone.

them. It's not turning on the pathway, it's increasing the volume of signaling if it is already turned on. It's not causing aggression, it's exaggerating the preexisting pattern of it, exaggerating the response to environmental triggers of aggression.

This transcends issues of testosterone and aggression. In every generation, it is the duty of behavioral biologists to try to teach this critical point, one that seems a maddening cliché once you get it. You take that hoary old dichotomy between nature and nurture, between biological influences and environmental influences, between intrinsic factors and extrinsic ones, and, the vast majority of the time, regardless of which behavior you are thinking about and what underlying biology you are studying, the dichotomy is a sham. No biology. No environment. Just the interaction between the two.

Do you want to know how important environment and experience are in understanding testosterone and aggression? Look back at how the effects of castration were discussed earlier. There were statements like "Remove the source of testosterone in species after species and levels of aggression typically plummet." Not "Remove the source . . . and aggression always goes to zero." On the average it declines, but rarely to zero, and not at all in some individuals. And the more social experience an individual had being aggressive prior to castration, the more likely that behavior persists sans *cojones*. Social conditioning can more than make up for the hormone.

Another example, one from one of the stranger corners of the animal kingdom: If you want your assumptions about the nature of boy beasts and girl beasts challenged, check out the spotted hyena. These animals are fast becoming the darlings of endocrinologists, sociobiologists, gynecologists, and tabloid writers. Why? Because they have a wild sex-reversal system—females are more muscular and more aggressive than males and are socially dominant over them, rare traits in the mammalian world. And get this: females secrete more of certain testosterone-related hormones than the males do, producing the muscles, the aggression (and, as a reason for much of the gawking interest in these animals, wildly masculinized private parts that make it supremely difficult to tell the sex of a hyena). So this appears to be a strong vote for the causative powers of high androgen levels in aggression and social dominance. But that's not the whole answer. High up in the hills above the University of California at Berkeley is the world's largest colony of spotted hyenas, massive bone-crunching beasts who fight with each other for the chance to have their ears scratched by Laurence Frank, the zoologist who brought them over as infants from Kenya. Various scientists are studying their sex-reversal system. The female hyenas are bigger and more muscular than the males and have the same weirdo genitals and elevated androgen levels that their female cousins do back in the savannah. Everything is in place except . . . the social system is completely different from that in the wild. Despite being stoked on androgens, there is a very significant delay in the time it takes for the females to begin socially dominating the males—they're growing up without the established social system to learn from.

When people first grasp the extent to which biology has something to do with behavior, even subtle, complex, human behavior, there is often an initial evangelical enthusiasm of the convert, a massive placing of faith in the biological components of the story. And this enthusiasm is typically of a fairly reductive type—because of physics envy, because reductionism is so impressive, because it would be so nice if there were a single gene or hormone or neurotransmitter or part of the brain that was *it,* the cause, the explanation of everything. And the trouble with testosterone is that people tend to think this way in an arena that really matters.

This is no mere academic concern. We are a fine species

with some potential. Yet we are racked by sickening amounts of violence. Unless we are hermits, we feel the threat of it, often as a daily shadow. And regardless of where we hide, should our leaders push the button, we will all be lost in a final global violence. But as we try to understand and wrestle with this feature of our sociality, it is critical to remember the limits of the biology. Testosterone is never going to tell us much about the suburban teenager who, in his after-school chess club, has developed a particularly aggressive style with his bishops. And it certainly isn't going to tell us much about the teenager in some inner-city hellhole who has taken to mugging people. "Testosterone equals aggression" is inadequate for those who would offer a simple solution to the violent male—just decrease levels of those pesky steroids. And "testosterone equals aggression" is certainly inadequate for those who would offer a simple excuse: Boys will be boys and certain things in nature are inevitable. Violence is more complex than a single hormone. This is endocrinology for the bleeding heart liberal—our behavioral biology is usually meaningless outside the context of the social factors and environment in which it occurs.

FURTHER READING

For a good general review of the subject, see E. Monaghan and S. Glickman, "Hormones and Aggressive Behavior," in J. Becker, M. Breedlove, and D. Crews, eds., *Behavioral Endocrinology* (Cambridge, Mass.: MIT Press, 1992), 261. This also has an overview of the hyena social system, as Glickman heads the study of the Berkeley hyenas. For technical papers on the acquisition of the female dominance in hyenas, see S. Jenks, M. Weldele, L. Frank, and S. Glickman, "Acquisition of Matrilineal Rank in Captive Spotted Hyenas: Emergence of a Natural Social System in Peer-Reared Animals and Their Offspring," *Animal Behavior* 50 (1995): 893; and L. Frank, S. Glickman, and C. Zabel, "Ontogeny of Female Dominance in the Spotted Hyaena: Perspectives from Nature and Captivity," in P. Jewell and G. Maloiy, eds., "The Biology of Large African Mammals in Their Environment," *Symposium of the Zoological Society of London* 61 (1989): 127.

I have emphasized that while testosterone levels in the normal range do not have much to do with aggression, a massive elevation of exposure, as would be seen in anabolic steroid abusers, does usually increase aggression. For a recent study in which even elevating into that range (approximately five times normal level) still had no effect on mood or behavior, see S. Bhasin, T. Storer, N. Berman, and colleagues, "The Effects of Supraphysiologic Doses of Testosterone on Muscle Size and Strength in Normal Men," *New England Journal of Medicine* 335 (1996): 1.

The study showing that raising testosterone levels in the middle-ranking monkey exaggerates preexisting patterns of aggression can be found in A. Dixson and J. Herbert, "Testosterone, Aggressive Behavior and Dominance Rank in Captive Adult Male Talapoin Monkeys *(Miopithecus talapoin)*," *Physiology and Behavior* 18 (1977): 539. For the demonstration that testosterone shortens the resting period between action potentials in neurons, see K. Kendrick and R. Drewett, "Testosterone Reduces Refractory Period of Stria Terminalis Neurons in the Rat Brain," *Science* 204 (1979): 877.

The Graying of the Troop

Bill Martin, Transitions, *1978;*
courtesy Joseph Chowning Gallery, San Francisco

It was one of those days when all his joints ached. The rains had started, and the humidity inflamed his hips and knees, especially the one he had injured long ago in his youth. He hobbled across the field, off to find something to eat and then, perhaps, to sit with a friend. Foraging for palatable food had grown more difficult as he lost more of his teeth—sometimes he spent half the day trying to appease his hunger. He paused for a moment, a bit disoriented, briefly unsure of his direction. When he was young, he had sprinted and chased and wrestled with his friends in this field.

As he resumed on the path, the two of them emerged from behind a tree, coming toward him. He tensed. They were new to the area, strangers, but not quiet like new ones usually were. Instead they were confident, making trouble for everyone. They worked together. Maybe they were brothers.

He was frightened, breathing faster. He hesitated for an instant to scratch nervously at a sudden itch on his shoulder. He glanced down, making no eye contact, as they approached, and felt a swelling of relief as they passed him. It was short-lived, as they suddenly spun around. The nearest one shoved him hard. He tried to brace himself with his bad leg but it simply wasn't strong enough, and it buckled under him. He shrieked in fear. They leapt on him, he saw a flash of something sharp, and, in a spasm of pain, felt himself cut, slashed across his cheek and then from behind, on his back. It was over in a second and, as they ran off, leaving him on the ground, the second one yanked painfully on his tail.

As discussed in a number of earlier essays, for nearly twenty years, I have spent my summers studying the behavior and physiology of a population of wild baboons living in the grasslands of East Africa. These are smart animals with pungent, individualistic personalities, living out twenty- to twenty-five-year life spans in large social groups. I've been able to experience the shock of mortality in a truncated way with them—watching baboons who used to be terrors of the savannah get hobbled with arthritis, now lagging behind when the troop forages for food, or finding myself shuddering at the weathered state of a male, when he and I were both subadult primates in the 1970s when I began my work. They've gotten old on me.

With the graying of some of my troop, I've gotten the chance to understand how the quality of their later years reflects the quality of how they lived their lives. What I've seen is that elderly male baboons occasionally carry out a behavior that is initially puzzling, inexplicable. The reason why it occurs, as well as who carries it out, contains a lesson, I think, about how the patterns of a lifetime can come home to roost.

As discussed in "The Young and the Reckless," most Old World primate societies, such as those of baboons, are matrilocal—females spend their entire lives in the troop into which they were born, surrounded with female relatives. Males, on the other hand, tend to leave their "natal" troop around puberty, shipping off to parts unknown, off to make their fortune in a different troop. It is a pattern common to most social species—one of the sexes has to emigrate to avoid inbreeding. And, as discussed in that essay, an odd feature of such transfers among primates is that it is voluntary, unlike in many species where males are forced out by the resident breeding male at their first sign of secondary sexual characteristics, the first sign that they are turning into sexual competitors. Primates leave voluntarily; they get this itch, a profoundly primate wanderlust that makes it feel irresistible to be anywhere that is not the drab familiar home ground. They leave, gradually make their way into a new troop, are peripheral, subordinate, unconnected, and then slowly form connections and grow into adulthood. It is a time of life fraught with danger and potential and fear and excitement.

Occasionally, a prime-aged male baboon will change troops as well. Those cases are purely career moves—some imposing competitor has just seized the number one position in the hierarchy and is likely to be there for a while, a cooperative coalition of males has just unseated you from your own position, some such strategic setback occurs—now is an expedient time to try rising on some other troop's corporate ladder.

But what I and others have seen is that some elderly male primates will also transfer troops. This initially makes little sense, given what aging is about. This is no time of life for an animal to spend alone, unprotected in the savannah, during the limbo period of having left one troop but not yet having been assimilated into another. Senses are less acute, muscles are less willing, and predators lurk everywhere. The mortality risk for a young baboon increases two- to tenfold during the vulnerable transfer period; just imagine what it must be like for an elderly animal.

The oddity of senescent transfers becomes even more marked when considering what life will be like if the transition to the new troop is made safely. Moving into a world of strangers, of new social and ecological rules—which big, strapping males are unpredictably violent, who can be relaxed around? Which grove of trees is most likely to be fruiting when the dry season is at its worst? The new demands are made even harder given the nature of cognitive aging—"crystallized" knowledge, the recalling of facts and their application in usual, habitual ways, typically remains intact into old age, whereas declines are more common in "fluid" knowledge, the absorbing of new information and its novel and improvisatory application. It is an inauspicious time of life to try to learn new tricks.

The physical vulnerability, the emphasis on continuity, the reliance on the familiar. All this suggests that it is madness for a baboon to pick up in his old age and try a new life. Why should he ever do it?

A number of scientists, noting this odd event, have come up with some ideas. Perhaps the grass is greener in some other troop's field, making for easier foraging. Maybe the male believes that a move is just the thing to rejuvenate his career, the chance for a brief last hurrah in some other troop. Some evidence for this idea has been found by Maria van Noordwijk and Carel van Schaik, of the University of Utrecht in the Netherlands, in their studies of macaque monkeys. One particular theory has the poignant appeal of a Hallmark card— perhaps aged males leave in order to return to their natal troop, their hometown, in order to spend their final frail years amid the care and support of their aged sisters and female relatives. As another theory, aged males may leave the troop in which they were in their prime when their daughters become of reproductive age. This would serve as a means to keep aged males from inadvertently breeding with their daughters. Susan Alberts and Jeanne Altmann of the University of Chicago have found some indirect support for this hypothesis in their baboon studies. A particularly odd twist on this idea comes from Alison Richards of Yale University in her studies of sifakas, a Madagascan primate. In a mixture of King Lear and bad Marlin Perkins, it appears to *be* the reproductive-age daughters who drive out the elderly fathers. No Hallmark cards there for Father's Day.

The data from my baboons suggest an additional reason for the occasional transfer of an aged male: in many ways, the last thing you ever want to do is spend your older years in the same troop in which you were in your prime. Because the current dominant cohort treats you horribly.

When you examine the pattern of dominance interactions among the individuals in a troop, a distinctive pattern emerges. High-ranking males have their tense interactions with each other, one forcing another to give up a piece of food, a resting spot, disrupting grooming or sexual behavior. Number 3 in the hierarchy would typically have his most frequent interactions with number 2—breathing down his neck, hoping to slip in the knockout punch that ratchets him up a step in the pecking order—and with number 4, who is trying to do the same to him. Meanwhile, these high-ranking tusslers rarely bother with lowly number 20, a puny adolescent who has recently transferred into the troop (unless they are having a particularly bad day and need someone small to take it out on).

This is the general pattern regarding the frequency of dominance interactions. Inspect the matrix of interactions, though, and an odd pattern will jump out—the high-ranking males are having a zillion dominance interactions with number 14. Who's he, what's the big deal, why is his nose being rubbed in his subordinance at such a high rate? And it turns out that elderly number 14 used to be number 1, back when he was in his prime and the current dominant generation were squirrelly adolescents. They remember. When I examined the rare males who had transferred into my troops in their old age, they all experienced the same new world—they were utterly subordinate, sitting in the cellar of the hierarchy, but at least they were anonymous and ignored. In contrast, the males who declined in the troop in which they were in their prime were subjected to more than twice the number of dominance interactions as were the old anonymous males who had transferred in.

Why are the prime-aged males so awful and aggressive to the deposed ruling class? Are they still afraid of them, afraid that they may rise again? Do they get an ugly, visceral thrill out of being able to get away with it? It is impossible to infer what is going on in their heads. My data indicate, however, that the new generation doesn't particularly care *who* the aged animal is, so long as he was once high-ranking. In theory, one might have expected a certain grim justice of what goes around comes around—males who were particularly brutal in their prime, treating young subordinates particularly aggressively, would be the ones most subject to the indignities of old age when time turned the tables. But I did not observe that pattern—the brutalizing of elderly males who remained in the troop was independent of how they went about being dominant back when. All that seemed to matter was that they once were scary and dominant. And that they no longer were.

This all seems rather sad to me, this feature of life among our close relatives—how, in one's old age, it can be preferable to take your chances with the lions than with the members of your own species, how one might choose to rely on the kindness, or at least the indifference, of strangers.

Looking at this grim pattern, the puzzlement before—why should an elderly male ever leave the troop?—gives rise to the inverse—why should he ever stay? Yet roughly half of them do, living out their final years in the troop in which they were in their prime. Are these the ones who have held on to a protective high rank longer, who for some reason are less subject to the harassments of the next generations? My data indicated that this is not the case. Instead, the males who stayed had something unique to sustain them during the hard times—friendship.

Aging, it has been said, is often the time of life spent among strangers. Given the demographics of aging among humans, old age for a woman is often spent among strangers because she has outlived her husband. And given the typical patterns of socialization that have been documented, old age for a man in our society is often spent among strangers because he has outlived his perceived role—job-holder, bread-winner, careerist—and has discovered that he never made any true intimates along the way. Studies of gender-specific patterns of socialization have shown how much more readily women make friends than do men—they are better able to communicate in general and about emotions in particular, more apt to view cooperation and interdependence as a goal rather than as a sign of weakness, more interested in emotionally affirming each other's problems than simply problem solving. And by old age, women and men differ dramatically in the number of friends and intimates they still have. In an interesting extension of this, Teresa Seeman and colleagues at Yale University have shown that close friendships for an older man are physiologically protective, while for a woman, it is the quality of the friendship that is important. Interpretation—intimate friendships are so rare for an older man that that in and of itself is a unique and salutary marker. In contrast, for an aged woman, intimate friends are the norm, not such a big deal in and of itself. Far too often, intimacy at that age means the woman has the enormously stressful task of taking care of someone infirm, hardly a recipe for her own successful aging. Seeman's studies show physiological protection for the women whose intimate relationships are symmetric and reciprocal, rather than a burden.

I think there are some parallels here to the world of baboons. Females, by dint of spending their whole lives in the same troop, are surrounded by relatives and by nonrelatives with whom they've had decades to develop relationships. Moreover, social rank in females is hereditary and, for the most part, static over the lifetime, so the jostling and maneu-vering for higher rank is not a feature in the life of a female. In contrast, a male spends his adult years in a place without relatives, typically apart from the individuals he grew up with, where the primary focus of his social interactions is often male-male competition. In such a world, it is a rare male who has friends.

To use this term is not an anthropomorphism. Male baboons do not have other males as friends—for example, over the years, I can count on one hand the number of times I've seen adult males grooming each other. The most a male can typically hope for out of another adult male is a temporary business partner in a coalition, and often an uneasy partner at that. For the rare male who has "friendships," it is with females.

This has nothing to do with sex—shame on you for such tawdry, lascivious thoughts. It's purely platonic, independent of where the female is in her reproductive cycle. They're just friends. This is an individual that the male grooms a lot, and who grooms him in return. It is someone he will sit in contact with when either is troubled by some tumult in the baboon world. It is someone whose infant he will play with or carry protectively when a predator lurks.*

Barbara Smuts, of the University of Michigan, published a

---

*And if that's not friendship, what is? This is not Disney cutesying of animals—instead, these patterns are discerned with rigorous, quantitative data by some of the best primatologists around. Moreover, it is not anthropomorphic to use terms like "personality," "community," or even "culture" when discussing primates either. As but one example of this (and some references are given at the end of this piece for more reading on these subjects), chimpanzees in the wild are now known to make a variety of tools, an observation in and of itself that has to be among the most provocative scientific findings of the century. Primatologists are now observing that there are differences in different chimp populations in the style of tool construction and use. There's not some sort of hard-wired instinctual drive in chimps to construct hedge trimmers for use in the rain forest. Instead, they learn, experiment, improve with their tools in a way that reflects regional differences in tool-making cultures.

superb monograph a decade ago analyzing the rewards and heartbreaks of such friendships, trying to make sense of which of the rare males are capable of such stability. And she documented something that I know many baboonologists have observed in their animals: males who develop these friendships are ones who have placed a high priority on them throughout their prime adult years. These are males who would put more effort into forming affiliative friendships with females than making strategic fighting coalitions with other males. These are the baboons who maximize reproductive success through covert matings with females who prefer them, rather than through the overt matings that are the rewards of successful male-male conflict. These are males who, in the prime of life, might even have walked away from high rank, voluntarily relinquishing dominant positions, rather than having to be decisively defeated (and possibly crippled) in their Waterloo.

Work by Smuts, Craig Packer of the University of Minnesota, and Fred Bercovitch of the Caribbean Primate Center has shown that male baboons are more likely to form such affiliative relationships with females as they mellow into old age. But, to infest the world of baboons with some psychobabble, the males with the highest rates of these affiliative behaviors are the ones who made their distinctive lifestyle choices early on. And it is this prioritizing that differentiates them in their old age. When I compared males who, in their later years, remained in the same troop with those who left, the former were the ones with the long-standing female friendships—still mating, grooming, being groomed, sitting in contact with females, interacting with infants. These are the males who have worked early on to become part of a community.

A current emphasis in gerontology is on "successful aging," the study of the surprisingly large subset of individuals who spend their later years healthy, satisfied, and productive. This is a pleasing antidote to the view of aging as nothing but the dying of the light. Of course, how successfully an individual ages can depend on the luck of the draw when it comes to genes, or the good fortune of the right socioeconomic status. Yet gerontologists now appreciate how much such successful aging also reflects how you live your daily life, long before reaching the threshold of old age. It's not a realm where there's much help from quick medical fixes or spates of resolutions about eating right/relaxing more/getting some exercise, starting first thing *tomorrow.* It is the despair of many health care professionals how difficult humans find the sorts of small, incremental daily acts that constitute good preventative medicine. It looks like when it comes to taking the small steps toward building lifelong affiliations, your average male baboon isn't very good at this either. But the consequences for those who can appear to be considerable; there's a world of difference between doing things today and planning to do things tomorrow.

No doubt, somewhere, there is a warehouse crammed with unsold, musty merchandise, left over when various fads of the sixties faded, cartons of cranberry-striped bell-bottoms and love beads. There may still be a market for some of that stuff: perhaps baboons might pay attention to one of those posters that we all habituated to, the horribly clichéd one about today being the first day of the rest of your life.

FURTHER READING

For a more technical treatment of this subject, see R. Sapolsky, "Why Should an Aged Male Baboon Ever Transfer Troops?" *American Journal of Primatology* 39 (1996): 149. Also see C. Packer, "Male Dominance and Reproductive Activity in *Papio anubis,*" *Animal Behaviour* 27 (1979): 37. Also see F. Bercovitch, "Coalition, Cooperation and Reproductive Tactics among Adult Male Baboons," *Animal Behaviour* 36 (1988): 1198.

A review of crystalline and fluid knowledge with respect to aging can be found in J. Birren and K. Schaie, *Handbook of the Psychology of Aging,* 3rd ed. (San Diego, Calif.: Academic Press, 1990).

The work of Maria van Noordwijk and Carel van Schaik can be found in "Male Careers in Sumatran Long-Tailed Macaques *(Macaca fascicularis),*" *Behavior* 107 (1988): 25. The studies of Susan Alberts and Jeanne Altmann are published in "Balancing Costs and Opportunities: Dispersal in Male Baboons," *American Naturalist* 145 (1995): 279. That paper also demonstrates the two- to tenfold increase in mortality when young baboons transfer. The work on sifakas is reported in A. Richards, P. Rakotomanga, and M. Schwartz, "Dispersal by *Propithecus verreauxi* at Beza Mahafaly, Madagascar: 1984–1991," *American Journal of Primatology* 30 (1993): 1.

Gender differences in patterns of successful aging is the subject of T. Seeman, L. Berkman, D. Blazer, and J. Rowe, "Social Ties and Support and Neuroendocrine Function: The MacArthur Studies of Successful Aging," *Annals of Behavioral Medicine* 16 (1994): 95. For a wonderful overview of gender differences in emotional expressivity, see Deborah Tannen's 1990 book, *You Just Don't Understand* (New York: Morrow). I firmly believe this should be required reading for all newlyweds.

Antidotes to worries about anthropomorphisms about primates: For the definitive discussion of platonic intimacy, see B. Smuts, *Sex and Friendship in Baboons* (New York: Aldine Publishing Company, 1985). For temperament and personality among primates, see A. Clarke and S. Boinski, "Temperament in Nonhuman Primates," *American Journal of Primatology* 376 (1995): 103. Also see J. Ray and R. Sapolsky, "Styles of Male Social Behavior and Their Endocrine Correlates among High-ranking Baboons," *American Journal of Primatology* 28 (1992): 231. And for an overview of cultural differences among primates, see R. Wrangham, *Chimpanzee Cultures* (Cambridge, Mass.: Harvard University Press, 1994).

The general and encouraging subject of aging not always being dreadful is reviewed in P. and M. Baltes, *Successful Aging* (Cambridge, Eng.: Cambridge University Press, 1990).

Curious George's Pharmacy

Mark Tansey, Nature's Ape, 1984;
private collection, Honolulu, courtesy Curt Marcus Gallery, New York

It was the first lecture of a neurobiology course in graduate school, and the instructor did a fabulous job of tripping us up on a romantic bias. The lecture was about synapses, the tiny gaps in between cells in the brain. The lights came down and, unexpectedly, the first slide was of a beautiful Egyptian tablet, filled with hieroglyphics. We hushed as he started off in a stagy Rod Serling voice. "During the reign of Amenhotep the First, more than 3,500 years ago, at the height of their power, the ancient Egyptians . . . knew nothing whatsoever about the synapse." We all fell for it, expecting instead to have been told about how detailed descriptions of the synapse could already be found in that fragile papyrus. It's the Wisdom of the Ancients syndrome, the public TV phenomenon where no fact in some field like astronomy, for example, can be cited without first noting how the Egyptians, Greeks, or Incans knew that centuries ago. Brain surgery, flying machines, how to get the VCR to record automatically—they knew it all back then. Then there's the Wisdom of the Non-Westernized syndrome, a relative newcomer, but an irresistible one for the folks who'd love to go dancing with wolves. And for a crowd-pleaser, there's always the Wisdom of the Children.

But if you want a real tried-and-true performer, go with the Wisdom of the Animals. For us humans, our well-advertised intelligence so rarely seems to be user-friendly; it is such a taxing, muscular process. We rack our brains, sort out our options, put on our thinking caps. It's all so, well, unnatural. "Brainstorming" even sounds neurologically dangerous, as if our neurons might short-circuit. But animals, on the other hand, just *know*. No energetic striving for them; theirs is the realm of instinct and reflex. They fly vast distances without getting lost, unerringly find food they've buried, recognize relatives on first encounter. They sense impending earthquakes, intuit human emotions, predict downward trends in the stock market. It is a seamless kind of wisdom that reflects a profound equilibrium with the environment.

Much of this is nonsense. Animals sometimes do indeed perform amazing feats of intelligence. But the idea of *animaux savants* is soggy with romanticism. Animals are not merely machines of effortless cognitive instinct. Their intelligence is more of an active, exploratory process, nowhere more prominent than in primate behavior. In the name of social competition, primates scheme and plot in remarkably sophisticated and all too familiar ways: they think through their next gambit; they try out new strategies; they screw up and learn from their mistakes.

All of this is a preamble to evaluating a new version of the Wisdom of the Animals. It's one that has generated tremendous excitement and, I feel, not enough skepticism. So the goal here is to help the reader to ask, What exactly do these beasts know and how do they know it?

In the late 1970s the primatologists Richard Wrangham of Harvard University and Toshisada Nishida of Kyoto University, conducting fieldwork in the Gombe National Park in Tanzania, noticed something odd about the feeding patterns of the local chimpanzees. On waking, presumably with an empty stomach, some of the animals would dine on the leaves of a shrub called *Aspilia*. What was strange was the way they ate the leaves:

instead of chewing them, the chimpanzees wadded them under their tongues, held them there for a while, and then swallowed them whole. Off and on throughout the meal the chimpanzees would make faces: *Aspilia* apparently was no tasty treat. Stranger still, the leaves passed through the animals' guts undigested, emerging essentially intact in their feces. Unpleasant to the palate, with little nutritive value, *Aspilia* posed a question: Why did the apes even bother?

Intrigued, Wrangham sent samples of the plant for analysis to Eloy Rodriquez, a biochemist at the University of California in Irvine. Rodriquez discovered that the leaves contain thiarubrine-A, a reddish oil known to be a potent toxin against fungi, bacteria, and parasite nematodes. Indeed, he noted that the youngest *Aspilia* leaves, much the preferred kind among the chimpanzees, are far richer in the oil than the older leaves are. And he thought it significant that the apes kept the drug under their tongues before gulping it down: the tongue is dense with minute blood vessels, and so a substance held there can bypass the digestive juices and head straight into the circulation. (Physicians and victims of angina are well aware that nitroglycerin placed under the tongue can rapidly suppress chest pain.)

On the basis of his laboratory studies, plus Wrangham's observations, Rodriquez estimated that in each eccentric bout of *Aspilia* feeding, the chimpanzees consumed just enough thiarubrine-A to kill between 70 and 80 percent of the parasites in their digestive tracts, without harming other, useful intestinal bacteria. Wrangham and Rodriquez also noted that Tanzanian tribespeople have long chewed *Aspilia* leaves to heal stomachaches and a variety of other ills. Hence the two coworkers speculated that the chimpanzees were using the plant as medicine.

In the decade since that work was published, reports of self-medication by chimpanzees, and other primates, have become all the rage. Such is the vogue in academic circles that these studies travel under the imposing name zoopharmacognosy; a few years ago the American Association for the Advancement of Science even convened a symposium on the subject. And certainly, some compelling data were presented, including a report by the anthropologist Karen B. Strier of the University of Wisconsin in Madison. Strier has spent some ten years studying the Brazilian muriqui, or wooly spider monkey, one of many endangered species of the rain forest. Just before the onset of their mating season, Strier notes, muriquis seem hell-bent on reaching a rather inconvenient corner of their territory, where they consume large amounts of the legume *Enterolibium contortisiliqiuim*. As it happens, the fruit of the plant contains the steroid stigmasterol, a precursor of progesterone, the crucial female reproductive hormone. Thus, Strier speculates, consuming the legume helps bring on the mating season for the muriquis.

Apes and monkeys are far from being the only animals that appear insightful about eating foods for curing what ails them. There seems little doubt that creatures of every stripe have a host of strategies for detoxifying their diets, and in some cases the maneuvers are fairly complex. For example, in the endless wars between plants and animals, many plants have evolved leaves laced with secondary plant compounds, among which substances are cardiac toxins, hallucinogens, antifertility drugs, and growth inhibitors. Such chemicals have no effect on the plants, but they can be deadly poisonous to any animal foolish enough to eat them. Not to be outdone, animals have evolved counteroffensive patterns of behavior that enable them to eat the plants: they may follow up a poisonous meal by eating something that can detoxify the poison. Rats, for instance, often consume clay after eating highly toxic plants; the clay absorbs the poisons.

Animals also appear to correct their own dietary deficiencies.

In his famous "cafeteria" studies a half century ago the psychobiologist Curt P. Richter broke up a balanced rat diet into its constituent parts, serving up eleven small trays of proteins, oils, fats, sugars, salt, yeast, water, and so forth. Remarkably, the rats picked out an efficient diet that, with fewer calories, made them grow at a rate faster than that of rats fed normal chow.

Oh, those animals, and the wise and wonderful things they know. Wrangham, Rodriquez, and Strier, all leaders in the field, have been scrupulously careful in their reports and speculations. Nevertheless, the lay press has had a field day with these observations, nearly implying that somewhere out there, some chimp knows the cure for cancer. Despite the tremendous appeal of this new field, a surprising amount of work remains to be done before the most excited—and exciting—claims about zoopharmacognosy can be endorsed.

To what extent do animals' efforts to find pharmacologically active plants truly constitute self-medication? One step toward an answer is to ask another question: Does the purportedly medicinal food of animals really work? That issue is best illustrated if one imagines a reversal of roles between human and animal: A chimpanzee, for its science-fair projects, analyzes the remains in a potato chip bag discarded by a person in Topeka. Among the chemical constituents in the potato chips, the ape finds one compound that, when analyzed in a test tube, displays anticholera activity. Does that mean Topekans can prevent or cure cholera by eating potato chips? Analogously, is the amount of thiarubrine-A consumed by chimpanzees enough to kill the kinds of parasite that occur in the animals?

That question might seem to have been settled by Rodriquez, when he estimated that the amount of thiarubrine-A ingested by chimpanzees is enough to kill a large percentage of nematodes. But there is a catch: Rodriquez was working in the test-tube world of purified thiarubrine-A. In real life some of the drug could theoretically be inactivated by other compounds in the biochemically complex *Aspilia* leaf. And not all of the drug absorbed from under a chimpanzee's tongue is likely to reach its gut; some of the drug might be broken down first. In short, it must be shown experimentally that the quantity of *Aspilia* leaves consumed by the animal leads to fewer gut parasites. Such an experiment has not yet been reported.

Another issue is the role of *Aspilia* in the chimpanzee's overall diet. Is there really anything special about the plant, at least in its natural state? Is *Aspilia* the only plant near the Gombe reserve that contains thiarubrine-A? To put the issue into relief, if all the plants turn out to contain the compound, and if chimpanzees are thus eating thiarubrine-A all the time, why is a sick chimpanzee's selective diet any more exciting than, say, the fact that people eat more chicken than pigeon? Wrangham has not reported whether thiarubrine-A is unique to *Aspilia*. But Strier, in her work on muriquis, notes that most legumes in the rain forest inhabited by the monkey contain ample amounts of stigmasterol, making the animals' preference for *E. contortisiliqiuim* less impressive in that regard.

Now to a tougher question: Do animals go out of their way to consume certain foods when they are sick? Does a case of cholera in Topeka cause Topekans either to start eating potato chips or, at least, to step up their daily chip consumption? When a chimpanzee winds up with a gut full of parasites, does it suddenly get a craving for some *Aspilia* leaves?

The issue is critical. There is, as yet, nothing other than anecdotal evidence to suggest that the chimpanzees are showing intentionality. Intentionality, however, has been shown in one case, namely, the clay eating by rats I mentioned earlier. For example, after being fed the poison lithium chloride, rats eat clay to induce vomiting. If rats are conditioned to associate the taste of lithium chloride with saccharin, they will eat clay after eating saccharin without the lithium chloride. In other words, if a rat even thinks it has eaten a poison, it wants to eat clay. And that raises the most complex question of all: In such a valid case, how can the animals possibly know what cures their ailments?

The easiest response would be to fall back on the kind of explanation I mocked above: animals just know, instinctively, what to eat. Richter thought his rats balanced their diets by responding to instinctual cravings. Cravings do exist, and they are understood physiologically. When animals or people are hungry, for instance, they crave food, and as part of the craving the brain makes the olfactory receptors in the nose highly sensitive to the odors of food. The food becomes easier to find. Even more specific cravings have been demonstrated: When a rat is deprived of salt, it instinctively seeks out salt and finds the taste more pleasurable than usual. The neuroscientist Thomas R. Scott and his colleagues at the University of Delaware have shown that salt deprivation in rats causes neurons in the brain that normally respond to the taste of sugar to become responsive to salt. In other words, when a rat craves salt, the salt tastes as good to the rat's brain as sugar normally does.

The trouble with invoking innate cravings to explain cases of self-medication is that only two specific innate cravings have been demonstrated: one for salt and one for water. And so the plot thickens. If an animal does, in fact, pick out the correct medicinal food at the right time, it probably learned to do it. How could that happen?

Work by the psychologist Paul Rozin of the University of Pennsylvania throws some light on that question. Rozin wanted to study the building blocks of the nutritional expertise shown by Richter's rats. As a test, he chose to determine whether a rat compensates when it is suddenly deprived of a single important nutrient. He discovered that a rat deprived of, say, thiamine prefers a new thiamine-rich diet to its old thiamine-free one.

But what the rat learns is not that the new diet is good. Rather, the animal learns that the old one is bad. Animals are usually tentative about eating a new food; by only nibbling at it they are less likely to be seriously poisoned. But Rozin noted that animals with a thiamine-poor diet do quite the opposite. Given the choice, they enthusiastically dig into a new diet rich in thiamine—not because they sense the thiamine or its benefits but because anything is better than the thiamine-poor diet they have been getting. As proof, Rozin and his coworkers gave thiamine-starved rats a choice between their old thiamine-free diet and two new diets, one with thiamine and the other without it. In roughly equal proportions the rats all gravitated toward the two new menus, thereby getting, as a group, only half the thiamine replacement.

Those findings seem to explain what is taking place when a sick animal tries a new food or, more precisely, rejects its old one. But how does the animal figure out that the new food causes it to feel better? Timing is a problem. As much as a week can pass between the consumption of a new "medicinal" food and the onset of its beneficial effects. But according to classical learning theory, associations are best made when they are closely yoked in time. How could a chimp form an association between feeling better and the *Aspilia* leaves it first sampled a week before?

Work by the psychologist Martin E. P. Seligman, also of the University of Pennsylvania, may help explain the conundrum. According to Seligman, not all learned associations must be closely linked in time. He has named such delayed learning the "sauce béarnaise" syndrome, after an unpleasant experience with that rich food. Once, following a sumptuous dinner featuring a sauce béarnaise, Seligman spent an evening listening to Wagner at the opera. Back home, after midnight, he suddenly developed a case of gastroenteritis. The next time he sat down to a dinner that included sauce béarnaise, its taste made him queasy.

Linking the stomach pains to the sauce béarnaise eaten six hours earlier made no sense according to the traditional canons of learning theory. Organisms are supposed to associate punishment (in this case, the extreme one of feeling ill) with the events that immediately precede it. The hours of operatic music were far more recent than the dinner; by all logic, Wagner, not the food, should have nauseated Seligman. But Seligman and others have noted that not all such associations are made with equal ease. Instead, to apply the principle to the case at hand, one is far more likely to link stomach pains with a gustatory event (the dinner) than with an auditory one (the opera), even if the gustatory stimulus took place much earlier than the auditory one.

The sauce béarnaise syndrome, as originally formulated, dealt with making associations about bad events. But it should operate just as readily for associations about feeling better. Thus an ailing animal might happen upon a curative diet or a medicinal plant, and its bad feelings might make it try something new. If the new food then works reasonably quickly to make the ailment go away, the animal could associate the new food with the relief of its ailment.

It is by no means clear, however, whether that style of learning can explain all the purported cases of animal self-medication—primates and their diets, rats and their clay eating, dogs and cats and their grass eating. Consider the reported cases of self-medication with fertility plants. Even the sauce béarnaise syndrome could not account for an association between pregnancy and a particular food. What woman can remember what she ate for breakfast two weeks before she either did or did not have her menstrual period?

Building on Rozin's work, Bennett G. Galef Jr. and Matthew Beck, who were both psychologists at McMaster University, may have provided some answers. They noted that many of Rozin's thiamine-deprived rats, left to their own devices, were not great at learning which foods are effective remedies. But the social dimension makes a critical difference. Galef and Beck observed that a rat is far more likely to learn to eat a curative food if the animal is surrounded by other rats that already prefer the food. Moreover, a rat strongly conditioned to avoid a given food becomes less phobic about the food if other rats are happily eating it. In animals, as in people, social observation plays a powerful role in learning what to eat.

Social learning may go far to explain the phenomenon of zoopharmacognosy. How many people independently figure out how antibiotics work? How many learn to build an internal combustion engine on their own? Instead, people generally learn how to use medicine or technology by watching others who are proficient at it. Perhaps some chimpanzee with a miserable stomachache, and with a metabolism that responded quickly to thiarubrine-A, stumbled onto a rich vein of the stuff in some *Aspilia* leaves. All that remained was for the rest of the group to mimic the discoverer. And both primates and rodents are very good at mimicking behavior.

Notwithstanding such speculation, it remains true that in only one of the cases I have discussed here—that of nauseated rats eating clay—have experiments shown that animals make a concerted effort to eat a medicinal food. Furthermore, in no putative case of primate self-medication have the animals been shown to eat enough of the compounds to cure their ills. Finally, if rigorous experiments in the future demonstrate cases of effective self-medication, some pretty fancy learning theories will probably be needed to explain how the animals pull it off.

Somewhere down the line—after some essential and difficult research has been done—it may turn out that all the reports of zoopharmacognosy constitute legitimate animal self-medication. I am fairly skeptical about that prospect, but if it comes to pass, it will be a delightful result. Practically speaking, it might lead to the discovery of some new medicine for people; maybe somewhere out there really is a chimp who knows the cure for cancer. And more fundamentally—despite my own occasional cynicism—it's hard not to be moved by evidence that other species can learn to be wise about the world.

But even if none of the reports of zoopharmacognosy turn out to be true, the work of Rozin, Galef, and Beck demonstrates that animals can be pretty wise about the world. They know that if something ain't broken, don't fix it, and if something does need fixing, the best course is to keep an eye on individuals who are more experienced. Above all, they know that if things aren't working, it's a good idea to keep an open mind for new solutions. Those are useful lessons for human beings to learn—even from animals that might not turn out to be trained pharmacists.

EPILOGUE

Not surprisingly, some of the individuals whose work I discussed in this piece were not at all pleased. I reproduce their letter published in *The Sciences* and my response.

To the Editor:

Robert Sapolsky's review of work on medicinal plant use by animals implies in the most charming manner that he's a lone skeptic looking on a field of suckers. Mr. Sapolsky dismisses completely "the myth that animals inherently know all sorts of wise and wonderful things." Guess what: so do we. Contra Mr. Sapolsky's implications, we know of no one working in zoopharmacognosy who suggests that animal self-medication has been proved, or that primates have innate knowledge of medicinal plants. But we say Nonsense! to the idea that zoopharmacognosy is inherently stupid. We agree with Mr. Sapolsky that any medicinal knowledge gained by primates is likely to be culturally transmitted. But we haven't decided exactly how the practice develops, because there is no experimental evidence.

Mr. Sapolsky thinks the psychological mechanisms underlying the use of medicinal plants by apes are exclusively concerned with learning. Perhaps. But it might also be true that physical condition influences the degree of aversion to toxins. The theoretical biologist Marjorie Profet, now at the University of Washington in Seattle, pulled together the evidence that aversion to toxins increases in women when they are pregnant. When Jane Goodall was dosing sick chimpanzees, she put tetracycline into bananas. The sick chimpanzees accepted it, but the healthy ones didn't. As the sick chimpanzees improved in health, they stopped accepting the laced bananas. Furthermore, one of us (Michael Huffman) has noted that chimpanzees with the worst parasite loads tend to chew the bitterest leaves. A health-mediated aversion to noxious toxins would rank as a fascinating example of the principle that perception can be influenced in adaptive ways by "inherent" mechanisms laid down through natural selection. We're not saying it's been established, just as we haven't said animal self-medication has been proved to occur. But if it's too early to know whether apes self-medicate, it's certainly too soon to have a closed mind about how they do it!

Mr. Sapolsky caricatures the evidence of the role of thiarubrine (the putative medicinal in *Aspilia*) in self-medication by equating it with finding a rare anticholera chemical in a bag of potato chips. That's not even a cheap shot: it's simply wrong. First of all, we eat chips because they taste good and because they provide nutrients. But nothing suggests that *Aspilia* tastes good or that the plant provides nutrients for chimpanzees. The leaves are swallowed whole; they pass through the gut with so little damage that it takes a scanning electron microscope to detect rupture of the surface cells and trichomes ("hairs"). *Aspilia* leaves could provide food only if they had extraordinarily high concentrations of soluble nutrients—and work by the anthropologist Nancy Lou Conklin of Harvard University shows they do not. In short, *Aspilia* leaves are nothing like potato chips. Ripe fruits, for a chimpanzee, are like potato chips. So are chewed leaves. But a whole leaf is like a chunk of bark, or a bit of old leather, or a page from Mr. Sapolsky's article. You don't expect anyone to swallow it!

And unlike the potato chip chemical, thiarubrine occurs in *Aspilia* in high concentrations. It's not widespread in nature, however. It is known to be present in several members of the sunflower family, such as *Ambrosia,* but its special properties weren't discovered until 1984, when it was found in the plant *Chaenactis douglasii,* used by native Canadians to treat skin sores.

Thiarubrine is an exceptionally potent drug. It kills a wide range of fungi, bacteria and nematodes at low concentrations as well as or better than commercial preparations do. Thiarubrine is antiviral, has been found to reduce some solid tumors and has inspired three patent applications for its medicinal potential. That's more than one can say for the average potato chip chemical.

Even if *Aspilia* leaves are used medicinally, however, experiments so far have failed to determine their chemical value. In fact, the botanist Neil Towers of the University of British Columbia and his colleagues, despite repeated effort, have failed to find thiarubrine in the leaves, though the chemical is always present in the root of *Aspilia.* There is a growing body of evidence that *Aspilia* might actually be swallowed for its physical properties. Two of us, Huffman and Richard Wrangham, report in a forthcoming article that all the African apes swallow leaves and that all the leaves swallowed have one unifying feature: rougher ("hairier") surfaces than typically occurs on leaves in ape habitats. Perhaps there is some unknown benefit to swallowing a rough leaf; or perhaps there are desirable chemicals in the trichomes, as there are in some plant families such as the Asteraceae, to which *Aspilia* belongs. The search for answers goes on.

Of course, as Mr. Sapolsky notes, the critical evidence of self-medication will be consumption of medicinal plants by sick animals that subsequently get well. For chimpanzees, the evidence is anecdotal but striking. Huffman has twice observed patently sick chimpanzees in the Mahale Mountains National Park, in Tanzania, selecting a common plant item that is used hardly ever (possibly never) by healthy chimpanzees—the pith of *Vernonia amygdalina,* another medicinal plant used by people in many parts of Africa. In each case, the chimpanzees meticulously removed the leaves and outer bark from several young shoots and chewed on the exposed pith, sucking out the bitter juice. The animal's rate of recovery was comparable to that of local people, the Tongewe, who use the plant as a treatment for similar symptoms. The chemistry of *V. amygdalina* is well known. There are two major classes of bioactive compounds, sesquiterpene lactones and steroid glucosides. As with thiarubrine, those compounds have demonstrated exceptional antiparasitic activity. And yes, Mr. Sapolsky, those drugs are rare in the chimpanzee environment. Chewing the *Vernonia* pith may well become the first proven case of chemical self-medication in nonhuman primates.

Far less is known about any direct effects of plants on primate fertility, but the possibilities are compelling. Take the exploitation of *Enterolibium contortisiliqiuim* fruits by muriquis, the largest and one of the rarest of South American monkeys. What drew the attention of another of us, Karen B. Strier, to *Enterolibium,* compared with other fruits in the muriquis' diet, was the special journeys they made to the edge of their forest to reach the *Enterolibium* fruits. Muriquis consume the fruits only at the onset of the mating season. Interestingly, the fruits contain high levels of stigmasterol, a substance with well-documented effects on fertility in sheep and humans. The effects of stigmasterol on muriqui reproduction are not yet known, but these observations now make them a topic of considerable interest.

Since the first publications of *Aspilia*'s properties a range of observations has shown that elephants, bears, coatis, and capuchins all use natural chemicals as something other than food. We can take such anecdotes and build them into evidence. In an era of shrinking biological resources, we don't think it's such a good idea to rely only on studies of rats. Let's keep an open mind about ways of exploring the natural world. Who knows, out of little *Aspilia* leaves big zoopharmacognosies might yet grow.

Richard Wrangham
Harvard University
Cambridge, Mass.

Michael Huffman
Kyoto University
Kyoto, Japan

Karen B. Strier
University of Wisconsin
Madison, Wisconsin

Eloy Rodriquez
University of California
Irvine, California

And my reply:

I am pleased with the letter of Richard Wrangham et al., in that a careful reading of their letter and my piece shows substantial agreement. The authors of the letter, who are leaders in zoopharmacognosy, and I concur that the instances they postulate as being animal self-medication have not yet been proved, nor have the explanatory mechanisms underlying any putative cases. I'm sure they also agree with me that the science writers and members of the lay public who jumped to the conclusion that their exciting findings constituted proof need to be gently taught the pieces still to be filled in. That was the aim of my review.

In considering the as yet unproved premise that animal self-medication exists, I made heavy use of the potato chip example. I can assure the authors that I don't believe potato chips and *Aspilia* taste the same. The point of that example was that for chimpanzees to prove that humans self-medicate with potato chips, or for humans to prove the same about chimpanzees and *Aspilia,* the same things must be demonstrated: First, the food must have enough of the putative medicine to be curative under the conditions consumed. As the authors note, that has not yet been shown for chimpanzees and *Aspilia.* Second, the food must be relatively unique in possessing such a quantity of the compound. The authors note that such is the case for *Aspilia* and thiarubrine; to my knowledge, this is their first report of that fact. Third, sick animals must go out of their way to consume that food. In their letter, the authors present pleasing but anecdotal reports that chimpanzees have done so. Thus evidence is beginning to accrue in favor of that case of animal self-medication. As I noted in my piece, I'm delighted to see this take place—once it is proved, I'm as taken with a romantic fact as the next person.

The second speculative area regards the mechanisms underlying any animal self-medication. In contrast to the authors' reading of my piece, I do not think that such mechanisms are "exclusively concerned with learning." As I stated, "Social learning may go far to explain the phenomenon." I am quite open to explanations that invoke more physiological mechanisms once anecdotal reports are translated into more experimental evidence.

Finally, as a fellow primatologist and field biologist who understands the difficulty of studying rare and spontaneous events in animals other than the laboratory rat, I am pleased that we all agree in our conclusions. Whether or not zoopharmacognosy provides us with new drugs already used by animals that are trained (or intuitive) pharmacists, there are numerous other animal skills to be learned about along the way that more than justify these studies.

In other words, we were being a bunch of academics having a fine time arguing with each other over something. Since then, progress has been made on a number of fronts in zoopharmacognosy toward meeting some of the criteria discussed. Huffman and colleagues have continued to observe chewing of the bitter pith of *Vernonia amygdalina* and have recently reported that such chewing by a chimp with flatulence and diarrhea was associated with a drop in the amount of parasites in the animal's feces over the next day; the active compound in the plant remains unclear. Wrangham has recently shown that the rates of leaf-swallowing increase among chimps during times of tapeworm infestation. Both Wrangham and Huffman have obtained evidence suggesting that it is not so much the chemical properties of some of these leaves but their

mechanical properties that might have medicinal effects—
specifically, it looks as if intestinal worms get trapped in the
rough hairy surface of the intact leaf as it moves through
the intestinal tract. And in a recent review speculating on the
mechanisms by which putative medicinal plant use may be
acquired, Huffman and Wrangham emphasize a route heavily
dependent on Rozin's and Galef's work—during times of illness, animals become less averse to novel foods, and social
observation and learning play a large role in the propagation
of any discoveries about medicinal plants. Obviously, more
work is needed on this fascinating subject and, just as obviously, a critical plug must be made at this point—none of this
is going to matter much if we keep destroying the rain forests.
No one, regardless of their species, can learn much pharma-
ceutical science without a pharmacy.

FURTHER READING

For general overviews of zoopharmacognosy, see E. Rodriquez and R.
Wrangham, "Zoopharmacognosy: The Use of Medicinal Plants by Ani-
mals," in K. Downum, J. Tomeo, and H. Stafford, eds., *Recent Advances in
Phytochemistry* (New York: Plenum Press, 1993), 27, 89; M. Huffman and
R. Wrangham, "Diversity of Medicinal Plant Use by Chimpanzees in the
Wild," in R. Wrangham, W. McGrew, F. de Wall, and P. Heltne, eds.,
Chimpanzee Cultures (Cambridge, Mass.: Harvard University Press, 1994),
129; K. Strier, "Menu for a Monkey," *Natural History,* February (1993):42;
and A. Gibbons, "Plants of the Apes," *Science* 225 (1992): 921. These con-
tain references to the more technical reports. Huffman's recent report of a
drop of parasite load after feeding on *Vernonia amygdalina* can be found in
M. Huffman, S. Gotoh, D. Izutsu, K. Koshimuizu, and M. Kalunde, "Further Observations on the Use of the Medicinal Plant *Vernonia amygdalina*
(Del) by a Wild Chimpanzee, Its Possible Effect on Parasite Load, and Its
Phytochemistry," *African Study Monographs* 14 (1993): 227. Wrangham's
recent report concerning the correlation of leaf-swallowing with tapeworm infestation is found in R. Wrangham, "Relationship of Chimpanzee

Leaf Swallowing to a Tapeworm Infection," *American Journal of Primatology* 37 (1996): 297. [The recent studies suggesting it is the mechanical, rather than chemical, nature of the leaves that is medicinal can be found in: E. Messner and R. Wrangham, "In vitro testing of the biological activity of *Rubia cordifolia* leaves on primate *Strongyloides* species," *Primates* 37 (1996): 105—and in M. Huffman, J. Page, H. Ohigashi, and S. Gotoh, "Leaf-swallowing by chimpanzees for the control of strongyle nematode infections: Behavioral, chemical, and parasitological evidence." Congress of the International Primatological Society, Abstract 566.]

References for additional work cited in this piece: T. Scott and G. Mark, "Feeding and Taste," *Progress in Neurobiology* 27 (1986): 293. P. Rozin, "Acquisition of Stable Food Preferences," *Nutrition Reviews* 48 (1990): 106; also see P. Rozin and T. Vollmecke, "Food Likes and Dislikes," *Annual Review of Nutrition* 6 (1986): 433. Rozin's reviews also discuss the classic cafeteria experiments by Curt Richter. The sauce béarnaise syndrome is discussed in M. Seligman, *Biological Boundaries of Learning* (New York: Appleton-Century-Crofts, 1972), 8. See also: B. Galef and M. Beck, "Aversive and Attractive Marking of Toxic and Safe Foods by Norway Rats," *Behavioral and Neural Biology* 43 (1984): 298; and B. Galef, "Direct and Indirect Behavioral Pathways to the Social Transmission of Food Avoidance," *Annals of the New York Academy of Sciences* 443 (1985): 203.

The Dangers of Fallen Soufflés in the Developing World

Michael Poulton, All States Are Full of Noise, *1995;*
courtesy the artist

Then there was the summer that I took Joseph Odhiombo's* blood pressure once a week and, as a result, ate like a hog.

I was spending that summer, as I had for years, living in a tent in the far corner of a game park in Kenya, conducting stress physiology research. Each week, I'd drive thirty miles to the park headquarters to get some gasoline, pick up my mail, have a soda at the tourist lodge. Joseph had recently returned to the lodge as its manager. We had met years before, when I was starting my work and he had an entry-level position in the lodge. Someone in the multinational hotel chain had spotted his competence and discipline, and Joseph had rocketed up the corporate ladder, being transferred here and there among their establishments, learning the ropes. And now he had returned triumphantly to his original lodge, the flagship of the hotel chain, its manager at the inordinately young age of forty.

We were delighted to encounter each other again. His success showed; in the previous decade, Joseph had grown plump, more than plump, a rotund man in that African status symbol, a tailored leisure suit. And he had a problem that he confessed to me on our first meeting—he was suffering from

---

*Naturally, not his real name. "Odhiombo" is a common East Africa surname, the rough equivalent of "Smith" or "Jones."

severe high blood pressure. His doctor in Nairobi was worried, as he hadn't quite come up with a drug yet that was working. Joseph was openly anxious as to whether his life was in danger. It somehow came out that I had a blood pressure gauge, and soon I had volunteered to take his pressure each week. And so evolved our ritual.

It was one that flourished into ornate detail, as befit his role as chief of this pseudovillage. I'd arrive to find the corridor outside his office filled with excited underlings, lodge employees hoping to have their blood pressure read during the interim when I waited for Joseph to receive me. These were typically Masai tribesmen, only a few years out of the bush, who knew and cared nothing about blood pressure or heart disease, but understood that something ceremonial was happening and wanted in on it. I'd take their blood pressure—these lean, hard laborers, the baggage carriers, the gardeners, the workmen who built the place with only hand tools in the equatorial sun. Their pressure would be low, their heart rates an astonishing fifty beats a minute. To each, I'd say the same: "You are going to live to be a very old man."

Halfway through the hoi polloi, I'd be summoned into the inner waiting room, where the indoor staffers, the guys starting up the management ladder, would be fretting, waiting to roll up the sleeves of their off-the-rack leisure suits for their readings. Their numbers would be up, nothing catastrophic, but definitely on the high side, higher readings than you'd want for someone their age—in other words, usually about the same level as my own blood pressure.

Finally, I'd be summoned into Joseph's office. He'd greet me warmly, prepare himself under the watchful eye of his two assistant managers, there to learn the proper way to have their blood pressure taken. Silence, anticipation. And always, it would be way too high. "Joseph . . .," I'd say, in a scolding tone. He'd deflate with worry; it still had not come down. He'd tell me hopefully about the new antihypertensive medication his doctor in Nairobi was hoping to give him; the doctor hadn't quite gotten permission to import it into the country yet, but he was sure this one was going to help. . . . And I'd begin the next phase of our weekly ritual. "Maybe you should do some exercise, perhaps jog around the lodge a bit," I'd tentatively venture. He'd giggle happily at the absurdity of this—"But I am the manager; I cannot be seen running around my lodge." "Well, maybe you should go on a diet, cut back a bit." "But what will the tourists think of the food here if the manager looks like a starving man?" I'd make a few other hopeless suggestions until Joseph's anxiety would be subordinated to his hunger. Enough of blood pressure, he'd announce, I must join him for lunch. He would lead me into the corridor to proclaim his blood pressure to the assembled staff, who would gasp with admiration that their leader could achieve such high numbers with his heart. And then, it being impossible to say no to his enthusiastic hospitality, the two of us would descend upon the all-you-could-eat banquet, gorging with abandon.

So went that summer. Years later, as I ponder Joseph's ill health, it strikes me that this was a man who became sick from being partially Westernized, and in a rather subtle way.

People in the developing world have became sick from being partially Westernized for a long time. Some cases represent the greatest of fears of well-meaning Western aid workers who plunge into some situation hoping to do good and instead make a mess of it. One example that has come back to haunt many a time: look at some nomadic tribe perpetually wandering around a howling wasteland of a desert searching for daily water for their goats and camels, the puddle of stagnant water here or there being all that saves them and their livestock. Obvious solution: haul in some equipment and drill a well for these folks. Expected outcome: no need to wander the desert anymore, plenty of water every day for the animals in this one convenient location. Unexpected additional outcome: the animals eat every leaf and tuft of grass within miles, and the high density of animals and their owners now encamped around the well makes communicable diseases leap through herds and through human populations with greater speed and virulence. Result: desertification, pandemics. You don't want Western centralization of resources until you have Western resistance to diseases of high population densities.

Sometimes people in the developing world became sick from partial Westernization because their bodies work differently from those of Western populations. Think about the world of subsistence farmers eking out a living on their few tropical acres, or the nomadic pastoralists wandering the equatorial grasslands or deserts with their herds. Food and water are often scarce, unpredictable. Salt, essential for life and found in low levels in most foods other than meat, is something to trade riches for, something to go to war over. Bodies make use of these treasures mighty efficiently. Have the first smidgen of sugar and carbohydrates from a meal hit the bloodstream and the pancreas pours out insulin, ensuring that every bit of those precious nutrients is stored by the body. Similarly, the kidneys are brilliant at reabsorbing salt and thus water back into the circulation so that nothing is wasted.

Scientists have known for decades that people in the developing world are particularly good at retaining salt and water, that they often have hair-trigger metabolisms that are markedly efficient at storing circulating sugars. These highly adaptive, evolved physiologic traits are called "thrifty genes." This represents a classic case of viewing the world through Western-colored glasses. It's not that people in the developing world have particularly thrifty metabolisms. They have the same normal metabolisms that our ancestors did, and that nonhuman primates have. The more apt description is that

Western surplus allows for the survival of individuals with sloppy genes and wasteful metabolisms.

Semantics aside, what happens when people in the developing world suddenly get access to Western diets, the newly emerging middle class of the Third World luxuriating in tap water and supermarkets stocked with processed sugars, a chicken in every pot and a saltshaker on every dinner table? Bodies that retain every bit of glucose and every pinch of salt now develop big problems. Throughout Africa, the successful middle class now suffers from a veritable plague of hypertension, hypertension known to be driven by renal water retention rather than by cardiovascular factors as is more often the case. And throughout the developing world, there are astronomically high rates of insulin-resistant, adult-onset diabetes, an archetypal Western disease of nutrient surplus—for example, among some Pacific islanders, Western diets have brought them ten times the diabetes rate found in the United States. There's a similar, if more complex, story having to do with non-Western patterns of stomach acid secretion as a defense against oral pathogens, with the price of a greatly increased risk of ulcers once you adopt a Western lifestyle. The punch line over and over—you don't want to receive the gift of a Western diet until you have a Westernized pancreas or kidneys or stomach lining.

Joseph and his hypertension seemed to fit this picture perfectly. But as I've thought more about him, I've become convinced that he suffered from an additional feature of partial Westernization. And this is one that I now recognize in a number of my African acquaintances—a lecturer at the university, nearly incapacitated by his stomach ulcer; an assistant warden in the game park, racked with colitis; a handful of other leisure-suited hypertensives. As a common theme, they all take their newly minted Western occupations and responsibilities immensely seriously in a place where nothing much ever works, and they haven't a clue how to cope.

Africa is a basket case, to the endless despair of those of us who have lived there and have grown to love it and its people. Find its corners that are still untouched by the maelstroms of this century, and villagers thrive there with wondrous skill and inventiveness. But much of the rest of it is a disaster—AIDS, chloroquine-resistant malaria, and protein malnutrition; deforestation and desertification; fuel shortages and power outages; intertribal bloodbaths, numbing shuffles of military dictatorships, corrupt officials in their Mercedes gliding through seething shantytowns; Western economic manipulation, toxic pesticides dumped there because they're too dangerous to be legal in the West, decades of homelands turned into battlegrounds for proxy superpower wars. In a setting such as that, who could possibly care that the soufflé has fallen? Yet Joseph did.

Running a lodge was a hopeless task. During peak tourist season, the front office in Nairobi would sell more rooms than existed and leave Joseph to deal with the irate tourists. The ancient generator on the freezer would expire, a spare part would be weeks away, and all the food would rot, leaving Joseph to negotiate with the local Masai to buy leathery old goats and curry them up beyond recognition for the evening dinner. The weekly petrol truck would be rolled off some bridge on the trip from Nairobi by its drunken driver, stranding all the lodge vehicles as the tourists lined up for their game drives. I'd seen plenty of other managers, before and after Joseph, hide in the curio shop, masquerade as a chef, dodge all responsibility when some major snafu occurred. But Joseph, instead, would patiently wade into the middle of each crisis, take responsibility, issue apologies, and sit worrying in his office late into the night, searching for ways to control the uncontrollable. As would my lecturer friend, trying to do his research amid no funding, the university constantly closed because of antigovernment protests. And as would the assistant warden, actually caring about the well-being of the animals while his boss was busy elephant poaching, the government going months without paying him, or his armed rangers shaking down the local populace for cash. A mountain of findings in health psychology has taught us that the same physical insult is far more likely to produce a stress-related disease if the person feels as if there is no sense of control, and no predictive information as to how long and how bad the stressor will be. And the task of trying to make the unworkable work constitutes a major occupational stressor.

It seems to me that these individuals festered in their stress-related diseases in part because their jobs, if taken seriously instead of as window dressing, were inordinately more stressful than are the equivalent ones in the West. But in addition, the truly damaging feature of their partial Westernization was that they lacked some critical means of coping. The keys to stress management involve not only gaining some sense of control or predictability in difficult situations, but also having outlets for frustration, social support, and affiliation. The frequent chaos around these men like Joseph often guaranteed little or no sense of control or predictability. And the other features were missing as well. There was no resorting to the stress-reducing outlets we take for granted.

Imagine an ambitious young turk on the fast track of some high-powered, competitive profession, working the corridors of power in the company's headquarters in New York or L.A., someone finding the job to be increasingly stressful and abusive. What advice would you give? Go talk to the boss about the problems that need to be corrected; if need be, go over the incompetent superior's head or make an anonymous suggestion about the changes needed. Drop out of the rat race and go back to the carpentry you've always been good at, or maybe answer that headhunter's phone call and interview with some other company in the business. Get a hobby, do some regular exercise to let the steam out. Make sure you have someone to talk to—a spouse, a significant other, a friend, a therapist, a circle of people in the same business who'd understand; just make sure you're not carrying this burden alone.

It's an obvious list that comes to mind readily for most of us, as if we learned it in some Type A's Outward Bound course for surviving in the canyons of the urban outback. And for the stressed white-collar professional in a place like Kenya, few of these coping mechanisms are ever considered, or would work if they were tried. Repeatedly, in conversations with Joseph and my other friends and acquaintances in his position, I would try to import some of those ideas, and slowly learned how inapplicable they were.

Businesses are run in a strictly hierarchical fashion owing much to the elder systems in many villages. (I once had a wonderfully silly conversation with a young bank officer with endless who's-on-first-what's-on-second confusion—Now let me get this straight, it's your *village* chief who has picked out a wife for you and who tells you when is the best time for the wedding, but it's your *bank* chief who tells you what you should feed the guests at the wedding, but it's your *village* chief who . . .). One does not tell the boss how to run his business, one does not go over someone's head; this is a work world without suggestion boxes. Exercise for its own stress-reducing sake is unheard of; Africa, a century behind us in the trappings of conspicuous consumption, demands that its successful men look like bloated nineteenth-century robber barons, instead of like our lean, mean sharks cutting deals on the racquetball court. Hobbies are unheard of as well; no hotel chain, worrying about its stressed executives, encourages them to take up watercolors or classical guitar.

The nouveau Westerner in the developing world also often lacks the stress relief that we feel in knowing that there are alternatives. You don't decide to drop out, accept the smaller paycheck but at least be happy—there's typically a village of relatives depending on your singular access to the cash economy. Furthermore, you can rarely pick up and work for the opposition—there's only a handful of hotel chains, only one university, only one game park service; the infrastructure is still too new and small to afford many alternatives to a bad situation. And intrinsic in this is another drawback—there is less likely to be a peer group to bitch and moan with after work as an outlet, as people often have situations that are one of a kind.

The isolation and lack of social support has an even more dramatic manifestation. In a place like Kenya, the individuals who have been successes and have entered the cash economy—typically men—usually work in the capital or in one of the cities or, for some specialized job, in some distant outpost like a game park. And their wives and kids are back farming the family land in their tribal area (this pattern is so extreme that it has been remarked that Africa has evolved a form of gender-specific classism, with men forming the urban proletariat, and women, the rural peasantry). The average man in the workforce sees his family perhaps one month a year on home leave. You'll meet men who, when asked where they live, will say the name of their tribal village, at the other end of the country, and add as an afterthought, "But I'm working now in Nairobi," the capital—something they've been doing for twenty years. They just happen to spend eleven months a year temporarily sleeping near their workplace—far from their loved ones, far from the members of their tribe (a point of identification that has no remote equivalent for even the most ethnocentric of us in the West), in a country with few telephones to phone home and no such thing as quick weekend shuttle flights. And to my knowledge, there is no Westernized institution in all of Kenya that would say, "Well, given that you are such a success in the stressful position we have put you in, we're going to come up with some bonus money for someone to farm your family land so that your wife and kids can live here with you." There simply isn't the mind-set for considering that solution.

Perhaps the most unsettling lack of a coping mechanism is that there is little that can be done when something bad and unfair happens, and that could happen at any moment. It was often sheer luck that launched someone successful into this new world in the first place—in a lodge, the brightest individual around might be some pot washer stuck back in the kitchen, a man with a frustrated brilliance whose poor family couldn't pay school fees past the first few years, while the manager wound up where he was because he was the nephew of the subchief of the right tribe; no one's heard of Horatio Alger yet. And many of the nouveau Westerners I've gotten to know—catapulted into the new world by luck—are diligent, capable, and meritorious. Nevertheless, those traits still won't prevent it from all being yanked away at a moment's notice—a superior develops a grudge, there's another shuffle in the government and a new tribe emerges ascendant, and the job is suddenly gone. And there are few unions to defend you, few grievance boards to appeal to, no unemployment insurance. There is never the solace of knowing that there is redress, and the ground is never solid. And so, with a table filled with food each meal, with a king's ransom of salt always available, and with worries that never subside and tomorrow always seeming uncertain, a price is eventually paid.

Strong and pragmatic words have been written about the malaise of the developing world. Some of the most trenchant and unsparing have been written about Africa by one of its own, the political scientist Ali Mazrui. He views the economic, societal-wide problems of Africa as one of partial Westernization: "We borrowed the profit motive [of the

West} but not the entrepreneurial spirit. We borrowed the acquisitive appetites of capitalism but not the creative risk-taking. We are at home with Western gadgets but are bewildered by Western workshops. We wear the wristwatch but refuse to watch it for the culture of punctuality. We have learnt to parade in display, but not to drill in discipline. The West's consumption patterns have arrived, but not necessarily the West's technique of production."

An author like V. S. Naipaul, another jaundiced son of the developing world, has written of the same process of partial Westernization on the individual level, his characters often lost in limbo between the new world they have not quite yet joined and the traditional one they can never return to.

And the scientists concerned with thrifty genes merely describe a more reductive version of the price of partial Westernization. To which I add my observation that you don't want a Western lifestyle until, along with your Westernized physiology, you have Westernized techniques of stress management as well.

I once listened to a group of musicians playing in the lowest level of the labyrinthine Times Square subway station in New York, Peruvian Indians in traditional clothes, playing gorgeous Andean music with their hats out for change. Maybe these guys had already become slick New Yorkers with a good street act, but it was impossible to hear their music and not to think of the wispy memories of mountains they must have had in their heads, down there in the bowels of the urban earth.

Many of us in this country are only a few generations removed from the immigrant ancestors who made the long journey to this new world. Most who made that move would term it a success—they survived, even thrived, or at least their kids did, and they were now free of the persecution that had caused them to flee their homes. But regardless of the successes and benefits, the immigrant experience often exacted a considerable price for that first generation, the ones with only partially acculturated minds and bodies, the ones who forever carried those wispy memories of their world left behind.

After that summer of our blood pressure luncheons, Joseph was transferred to managing a different hotel, and we lost track of each other. I recently received word that his hypertension had indeed killed him, at age forty-eight. I am sorry that he is gone, as he was a good man. I hope his children will be able to experience the Western world that he gained for them more healthfully than he could. Before him, I only knew about immigrants who weathered long boat journeys in steerage in order to arrive, dislocated, in a new world. But now I realize the extent to which it is possible to be an immigrant in a dislocating new world that has journeyed to you.

FURTHER READING

The subject of thrifty genes can be found in a piece by the originator of the concept: J. Neel, "The Thrifty Genotype Revisited," in J. Kobberling and R. Tattersall, eds., *The Genetics of Diabetes Mellitus* (London: Academic Press, 1982), 283. For a discussion of the hypertensive consequences of plentiful salt in the developing world, see J. Diamond, "The Saltshaker's Curse," *Natural History,* October 1991, 20.

An overview of the building blocks of coping and stress management can be found in chapters 10 and 13 of my book *Why Zebras Don't Get Ulcers: A Guide to Stress, Stress-Related Diseases, and Coping* (New York: W. H. Freeman, 1994).

The Ali Mazrui quote is from a BBC special called *The Africans*. Similar sentiments occur in his book by the same title (Boston: Little, Brown, 1986).

The Dissolution of Ego Boundaries and the Fit of My Father's Shirt

⁓∞⁓

Alfredo Castañeda, El Puerto de Vera Cruz, *1993;*
courtesy Mary-Anne Martin/Fine Art, New York

I've been wondering lately how many bodies a person needs, not a question that would normally interest me as a scientist. Yet this question has crept in, and I'm no longer sure of the answer.

In retrospect, I realize that I once watched someone inhabit two bodies. It was Stephen Hawking, the astrophysicist as much famed for his Lou Gehrig's disease as for his work. He came to my graduate school to lecture on the beginning and end of time. Or perhaps the point, which I could barely catch with my rudimentary physics, was that time had no beginning or end. It was a dozen years ago and he could still move his mouth a bit, generating a gargly, novocainized, incomprehensible voice.

We packed the auditorium, biochemists and physiologists and geneticists, attending a lecture we would not understand to bear witness to Hawking's body melted by disease and his mind knowing whether time began. The door at the rear of the stage opened to reveal the university's four physics professors carrying Hawking, facing backward in his wheelchair. None were young men, they panted visibly, but it seemed as if only they could carry out this task, that if you did not understand special relativity or the mathematics of time's arrow and touched his wheelchair, you would be vaporized.

His chair was placed center stage, still facing backward.

The professors fled, never turning their backs on him. A stillness descended, followed by a whirring sound. His electric wheelchair rotated toward us, and we were confronted with a shriveled, mummified husk staring from behind horn-rim glasses. We inhaled as one, as if to chant ritualistically in some dead language.

So we sat there, paralyzed by this mummy brain from the crypts, when a young guy in jeans and corduroy jacket and tousled blond mane entered the stage. He had an offhanded, rolling gait. He was sauntering on the same stage as Stephen Hawking. He looked like he had just rolled out of bed and was still preoccupied with whomever he had left there. He walked up behind Hawking, readjusted his hair, and shoved the wheelchair forward.

Jesus, Hawking is rolling to the edge of the stage. We're dumbstruck, watching this rock star assassin murder Stephen Hawking. As the mummy brain approaches his doom, at the last second, the kid stops him. He roughly positions Hawking toward the audience, turns to us, and in an arrogant English public school voice says, "You know, you're not in church."

The kid seats himself next to Hawking, gripping the microphone like he's doing Vegas. This, apparently, is to be Hawking's voice, the person who will interpret the strangulated raspings for us.

Hawking begins and is indeed incomprehensible. Muffled garglings, pauses suggesting sentence structure, and the arrogant voice translating. "Today, I will discuss some theories of mine concerning the beginning and end of time. In some cases, these theories have been experimentally confirmed. In others, there has not yet been sufficient time for their confirmation." We chuckle—cocky bastard—before we catch ourselves, How could you think that about Stephen Hawking?

Hawking lectures with painstaking slowness but utter clarity—time having no beginning, radio waves doing the impossible and escaping from black holes. Equations appear that we can follow, interlaced with a puckish, arrogant humor.

The kid is worse. As Hawking struggles, he sits bored, tossing his chalk in the air, focused on an attractive woman in the front row. He drops the chalk and forces the mummy through the arduous task of starting again. Once, he translates a sentence and Hawking becomes agitated—the kid got it wrong. Hawking repeats himself. Glowering, the kid mutters, "You're not making yourself understood."

Who the hell is this kid? Slowly, it dawns on us. Hawking, in his Cambridge chair once occupied by Newton, has this kid as his student. And his chosen voice. This must be one incredibly smart guy and, not coincidentally, a deep intimate. Something is shifting in our picture of the mummy brain. Part of the hagiography of Hawking is that before the illness, he was a showboat, ostentatiously functioning at half speed to barely finish some task before returning to a party, all with a dissolute brilliance. He must have been just like this kid. And it hits us: they *planned* this, the whole show—the careening wheelchair, the "You're not in church," the kid's insouciance—to desanctify. The conspiracy we sense is so intimate that it is not a question of Hawking having been like the kid, or of this kid speaking for him. Instead, he *is* the kid for this hour. The mummy brain is gone. Instead, we're listening to an entertaining lecture by a cocky Cambridge don. One who just happens to require inhabiting two bodies to pull it off.

Hawking's act was metaphor and theater and, one assumes, temporary. The possibility of prolonged dissociation between numbers of bodies and consciousnesses has not often interested neuroscientists. One exception, however, is with split-brain patients. While the brain is roughly symmetrical, its functions can be *lateralized*—the left and right sides subsume different tasks. The left hemisphere typically specializes in language, while the right excels in nonverbal spatial abilities, facial recognition, music. Popular knowledge of such lateralization has fostered absurd New Age ideas that people go about every conceivable behavior with differing "Left Brain" or "Right Brain" styles. Nonetheless, the science of lateralization is solid.

The two hemispheres communicate by a hefty cable of connections called the corpus callosum. In one type of epilepsy, a seizure can propagate a subsequent one on the mirrored part of the other hemisphere, back and forth across the corpus callosum. In the 1960s, surgical severing of the callosum was tried on such epileptics, halting the seizures. However, the person was left with two disconnected hemispheres. Roger Sperry, who won the Nobel Prize for this work, designed brilliant experiments to feed information to only one hemisphere, or differing information to each simultaneously. He showed that the hemispheres could function separately, and with differing analytical strengths. For example, an object could be shown to the visual field such that the information only entered the verbal hemisphere, and the person could then readily identify the picture. But when the same information was presented to the nonverbal hemisphere, the person couldn't even state that he had seen anything—yet could identify the object by touch.

Each hemisphere could learn, remember, reason, have opinions, initiate behavior, be self-aware, have a sense of time and of future, and generate emotions. This raised the messy possibility that there were now two individuals occupying that skull. At an extreme was the idea that this was not only the case for split-brain patients, but that we all normally consist of two separate individuals, yoked together by the corpus callosum. Prompted by this, the psychologist Julian Jaynes wrote the highly eccentric *Origin of Consciousness in the Breakdown of the Bicameral Mind.* He argued that a coherent sense of self, a well-bounded ego, developed only some 3,000 years ago. Before that, the brain was "bicameral" (i.e., two-chambered), with minimal integration of the two hemispheres. Instead, one hemisphere spoke, either metaphorically or literally, and the other hemisphere obeyed, attributing that voice to the gods. Jaynes claimed that the modern sense of ego represents a breakdown of that bicamerality, that schizophrenics remain bicameral, and supported his ideas with mountains of trivia from archeology, mythology, the classics, and the Bible. The near consensus among savants was that the book was dizzyingly erudite, stimulating, and ultimately loony.

Sperry rejected the notion that there were two individuals inside anyone's head, and most agreed. While split-brain patients could be manipulated into displaying two independent cognitive styles, the underlying opinions, memories, and emotions were the same. This was explained anatomically. Even if the corpus callosum was cut, more primordial, deeper structures of the brain, critical to emotion and physiological regulation, remained connected. Split-brain brains are not really split into two, but instead form a "Y." There might be two separate consciousnesses, one paying attention to a speaking voice, the other to background music, one navigating through town by remembering names of streets, the other by remembering a spatial map of the town's appearance—yet it is still the same individual. One body, one person.

A battle as to how many selves can reside within one body has often raged over the issue of multiple personality disorder. Different facets of our personality dominate in different settings, sufficiently so that we may act like "a different person" when with a boss rather than a subordinate, or with a woman instead of a man. But we are not literally different people. In individuals with multiple personality disorder, however, separate personalities take full control of the person's behavior at different times. Most mental health professionals agree that there are individuals in whom the different facets of personality are so different, so disjointed and dissociated, as to be diseased. These patients often have suffered horrific childhood abuse, and it is theorized that the compartmentalizing of the different personalities evolved as a protective strategy. What is wildly controversial is whether these nonoverlapping identities truly represent different personalities, how this works biologically, and how often this occurs.

At one extreme are clinicians who claim to have had hundreds of such patients (bringing up the inevitable jokes about billing each personality separately). They cite studies showing that when the personalities of their patients shift, so too do eyeglass or medication prescriptions. Some of these practitioners even glory in the multiplicity of personalities. Their goal is not to bring about integration into a single core personality, as that represents some sort of drab, monochromatic surrender (some blather about paternalistic monotheism always seems to pop up around this point in the discussion), but instead to enable the patient to make happy, productive use of the various personalities.

At the other extreme are clinicians who have apoplexy over this, claiming that a true multiple personality patient is seen once in a career, that the "different eyeglasses" stories are only stories. The patriarchs of psychiatry have generally taken this view. In the latest edition of psychiatry's bible, the *Diagnostic and Statistical Manual,* there have been careful changes concerning the disorder. Its diagnosis no longer involves the "existence" of multiple "personalities." Instead, the prerequisite is the "presence" of "dissociated identities" (and, in fact, the disease is now called "dissociative identity disorder"). In other words, the key is that the patient identifies herself in multiple ways—and the experts won't touch with a ten-foot pole whether those identities really constitute personalities. Moreover, to paraphrase a psychiatrist who helped make those changes, the point is not that these people might have *more* than one personality but that, when the pieces are put together, they really have *less* than one.

"Multiple personality" disorder and split-brain patients raise the possibility of fragmentation of the self. Far more plausible is the idea of the self occasionally making room for another. Freudians believe that this can occur and can reflect profound psychopathology.

In the face of failure or loss, we all mourn with a certain sadness and withdrawal. And just as surely as we all can get depressed, most of us heal. Even in the face of great tragedy, such as the loss of a loved one, we can usually find the means, eventually, to feel that it is not the end of the world. However, some of us, in the face of such loss, fall into prolonged and incapacitating sadness—"melancholia," to use Freud's terminology, or a major depression, as we would now call it. Along with the usual symptoms seen in someone mourning, the deeply depressed individual typically shows a self-hatred, claiming the death was their fault, wallows in guilt over ancient behaviors, and engages in punishing self-destructive behavior. "In [the healthy state of] mourning," Freud wrote, "it is the world which has become poor and empty; in melancholia it is the ego itself." Major depression is "aggression turned inward," as Freud's colleague Karl Abraham termed it.

Why should the mourning typical of most of us give way to incapacitating sadness and self-hatred in some? The root, Freud thought, lay in ambivalence—the dead individual was not only loved but also hated—and the person's response to that ambivalence. Critically, the depressive is left unconsciously angry after the loss—anger for being abandoned by the dead person, for their previous conflicts, for the impossibility now of ever resolving them. As such, the depressive identifies with, incorporates, internalizes features of the lost individual, carries them inside. This is no antiseptic homage.

The emotional adversary is gone, and there is no choice but to reconstruct the lost person internally, and then carry on the battle.

In this, Freud made a critical observation. The features of the lost individual that are internalized, the opinions and habits, are not just any, but those that were most hated. "If one listens patiently to a melancholic's many and various self-accusations, one cannot in the end avoid the impression that often the most violent of them are hardly at all applicable to the patient himself, but that with insignificant modifications they do fit someone else, someone whom the patient loves or has loved or should love." By carrying on the most hated traits, one can still argue ("You see, don't you hate when I do that, can you believe I put up with that for fifty years?") and, through the terrible pain of a major depression, punish oneself for arguing.

Thus, for the Freudian, the major depressive's ego boundaries dissolve to allow parts of the lost individual to be carried within. Freud thought this explained the greatest psychodynamic ill of all—suicide. The ego can never consent to its own destruction; suicide requires the internalization of the other so completely that the suicide is more akin to murder.

Thus, science has occasionally considered cases of the self-fragmenting (or withering sufficiently to make room for another), cases where more than one self might dwell in one body. There has been even less concern for the possibility of one self occupying more than a single body. And, all things considered, these musings haven't generated much certainty, in any modern scientific sense, as to what really constitutes a "self." None of this had ever been of much interest to me as a scientist until recently, when I was personally exposed to a breakdown of the boundaries of self. At first, I could fully explain it with my science, viewing its extreme manifestations as pathology.

As my father aged, he suffered cognitive problems secondary to neurological damage, to the point where he would often not know what decade it was, his location, the names of his grandchildren. Along with this, his ego boundaries began to dissolve. Gradually, he purloined bits of my life—I, his only son. There were similarities already. Long ago, he had done medical research, as I do now; he had been a professor, as am I. We had always shared tastes, styles, and temperaments. But as he aged, the details of our lives began to intertwine. When I moved to San Diego, his navy years, recounted endlessly, suddenly included stretches spent in San Diego, so that soon we were sharing the same opinions and anecdotes about the town, forty years apart. When I moved to San Francisco, his entry to the United States switched from New York's Ellis Island to San Francisco, his reported first view of America including a Golden Gate Bridge that did not yet exist in the year of his entry. Wherever I went to lecture, it would now be a university where he was once visiting faculty. His medical research, cut short by the Depression, had been in cancer biology, but now he was full of contentless memories of an interest in neurobiology, my own subject. I don't believe it was competitiveness, or a need for us to have more in common. The problem was too much in common already. He identified passionately with whatever achievements I might have, and we knew how much I was him without the bad luck of refugee status, world wars, and the Depression, privileged with the rewards that his obsessive hard work had provided me, topped by perhaps a half century of life ahead as his own shadows lengthened. And as the fog of his disorientation swept in, he needed someone else's stories and became less certain of where he ended and I began.

This felt more than a bit intrusive, but I was well defended with an armamentarium of labels and diagnoses and detached, condescending understanding, a world where something disturbing is housebroken by turning it into a lecture. ". . . As another feature of the demented patient, one occasionally sees . . ." At night, he'd wander the house, agitatedly reporting the presence of angry strangers, of long-dead colleagues. If things were that bad, come on, cut him some slack if he's also confused about which of the two of us was falling in love with Californian redwoods. You see, there's been some neurological damage.

His recent death knocked me off that diagnostic high horse, as it became me who developed problems with boundaries. It started manageably enough. I began to spout his sayings and mannerisms. This was not Freudian melancholia—while I was plenty sad, I wasn't clinically depressed, I wasn't awash in ambivalence and anger, and the behaviors of his that I seized hadn't irritated me for decades with an Oedipal itch. They were the insignificant quirks that had made him him, and with which I now festered. I arranged the utensils as he did, hummed a favorite Yiddish tune of his throughout the day, looked at the landscape as I never had but as he always did. Soon, I had forsaken wearing my blue flannel shirts in order to wear the blue flannel ones of his that I had brought back with me. I developed an interest in his profession—architecture— for the first time. I grew up amid his blueprints and drafting board, but had remained indifferent to the subject. Yet now, I found myself absentmindedly drawing floor plans of my apartment, or trying to understand three-point perspective, something he had tried to teach me without success.

This seemed reasonable. When I was younger, signs that I carried bits of him in me would have triggered bristly Oedipal denial, defensive nit-picking about subtle differences. But I could deal with a certain amount of this with equanimity, homage without the Freudian bile. But things took a troubling turn.

When I had spent a week of mourning at the family house,

I observed the magnitude of his final frailty, the bottles of nitroglycerin everywhere. I took one back to California and kept it close to me for weeks. I would make love to my wife, work out in the gym, attend a lecture, and always, the bottle would be nearby—on a nightstand, in a sweat-jacket pocket, amid my papers. There was a day when I briefly misplaced it, and everything stopped for an anxious search. It was not that I had lost a holy relic of his suffering, an object to show my children someday to teach them about the man they hadn't known. This was urgent; I felt vulnerable. Was my heart now diseased, or was it his diseased heart somewhere inside me that I now vigilantly stood by to medicate?

What the hell was this all about? I don't believe in God or gods, seraphs or angels, transmigrations or transmutations of souls. I don't believe in souls, for that matter, or in UFOs, with or without Elvis on board. Was it my sense of hard-assed individualism that was feeling unnerved by this intermingling, or was I feeling his hard-assed sense of individualism being unnerved?

The height of the confusion came a month later, with the final lecture for the class that I was teaching. My mother, exhausted with caring for him and finally persuaded to leave him with a nurse and come for a vacation, had earlier sat in on my class, and the students had cheered when I introduced her. They had done a good thing. Four days later, he was dead, classes were canceled, and afterward, many of the students had expressed warm, supportive thoughts. I had come to feel close to all four hundred of them. At the end of the final lecture, I thought to tell them something about what a spectacular lecturer my father had been, things I had learned from his teaching that might apply to their lives. I intended to subject them to a eulogy about him, but something became confused and soon, wearing his shirt, I was lecturing for him, offering the frail advice of an octogenarian.

I warned them, amid their plans to tackle difficult problems in life and to be useful and productive, that they should prepare for setbacks, for the realization that each commitment entailed turning their backs on so many other things—like knowing their children, for example. And this was not me speaking, still with a sheltered optimism about balancing parenting with the demands of science, but he with his weathered disappointments and the guilt and regret he expressed in his later years that he was always working when I was a boy. I told them that I knew they wanted to change the world but that they should prepare for the inconceivable—someday, they would become tired. At the end, wondering whether this much emotion was setting me up for one of his angina attacks, I said good-bye for him to an ocean of twenty-year-olds rippling with life and future. And that night I put away the nitroglycerin.

During that month, my head swam with unlikely disorders from my textbooks to explain this intermingling. A year later, safe again on my battlefield of individuation, that time has begun to make more sense. I feel sure that what I went through need not merit a diagnosis, and I don't think anymore that his own earlier confusion about the boundaries of the two of us really had anything to do with his neurological problems. It is a measure of the pathologic consequences of my training as a scientist that I saw pathology that was not there, and a measure of the poverty of our times that I could only feel as a brief flicker something intrinsic to our normal human experience.

I have occasionally been able to observe that flicker. As a benefit of my years spent working intermittently in East Africa, I have often accompanied African friends to their homes, some hamlet clinging to a mountainside. Invariably, it would be obvious that my friend had become almost as much of an outsider as I was. By definition, if I knew this person, he was the one who had left. The second son, the restless one, the one who had gotten some schooling, a far-off job among other tribespeople, the one stumbling over words in his mother tongue, returning with new ways and a white friend.

And it would always be the same sort of world that my friend had left. There would be one old man who was The Old Man, one damaged village idiot who wandered the slopes, one drunkard wife-beater. And at the friend's house, there would always be the older brother, the first son, the one who had stayed. He would sit next to his aging father—silent, inexpressive farmers, unsophisticated men who spoke no English or Swahili, who had been to the county seat once, but no farther. They would sit there, half amused and half puzzled by some hyperverbal story told about the big city by my friend, grunting in unison in the same phlegmatic way. It is a world without our Western frenzy for individuation. In the schools, Peace Corps teachers would find it maddening that kids won't give answers when they know them; for fear of shaming their peers, kids are embarrassed to stand out. "The tall sheath of wheat gets cut first," goes the saying. It is a world where no parent thinks, "I want better for my kids," where it is absurd to ask a child what he or she wants to grow up to be. No one would ever wonder if it's a bit unhealthy for a thirty-year-old to still be living near the parents, or view joining the family business as a worrisome lack of independence. In this tough world, you're lucky if you wind up farming the same land or raising your kids the same way as your parents did, if you, as with these older brothers, turn into your parent, your identities melding.

Thomas Mann, in his retelling of the Bible story in *Joseph and His Brothers,* captures this. He describes young Joseph hearing the stories of old Eliezer, his father's servant, stories told in the first person that were the experiences of *the* Eliezer, the servant of Abraham from the mythic past. Mann describes how it was perceived as normal "that the old man's ego was not quite clearly demarcated, that it opened at the back, as it were, and overflowed into spheres external to his own individuality both in space and in time; embodying in his own experience events which, remembered and related in the clear light of day, ought actually to have been put into the third person." This Eliezer was not sharing parts of his life with the mythic one. Instead, as he aged, he had *become* the mythic Eliezer, and the community had *expected* him to do so.

It is the triumph of archetypes. In every traditional community, there is an Eliezer, the freedman who has become the wise servant. There must be an Esau, the primal force of simplicity, there must be the fratricidal conflict of brothers over the blessing of the ailing father, there must be an Abraham, the Ur-Patriarch. These needs transcend individual rights to a bounded ego, and people would be named and raised to be the next incarnation. An Abraham would always live nine hundred years, because in each community there would be that Abraham—he would simply need to inhabit a new body now and then.

This is not Jaynes's view, where there is not yet a sense of self. That sense is simply not that important, subordinated to something bigger and tribal.

This primacy of continuity is something we no longer have, and to feel even its glimmer, as I did, requires an emotional crisis, perhaps coupled with some seasoning. My students usually come with ego boundaries like exoskeletons. Most have no use for religion, precedents, tradition, or ritual, or if they do want ritual, it is not vertical over time, but newly minted and shared horizontally within their age-group. The ones who I train to become scientists go at it like the warriors they should be, overturning existing knowledge and reigning paradigms, each discovery a murder of their scientific ancestors. And if I have trained them well, I must derive whatever satisfaction I can from the inevitability of becoming their Oedipal target someday. In many ways, competitive science represents the most jagged version of our Western model, in which torches are wrested, rather than passed, intergenerationally. These students are right on maturational schedule in believing that they have burst anew free of sources and that they can reinvent the world within themselves. And if they should happen to find themselves becoming confused as to where they end and somebody else begins, it is obvious that they are dealing with something scientifically, certifiably abnormal.

That is becoming less the case for me. I can still do without religion, but some ritual would be nice. There're other changes. I watch these damn kids sprint past me when I play soccer; I fumble for an answer when watching *Jeopardy!* that the high school contestants jump on. My beard is showing some white; my spine has probably started shrinking while my refractory periods have lengthened. Another few birthdays and it becomes prudent to have a physician regularly poke around my prostate. It slowly dawns that my ego-bounded self is not such a hot deal anymore.

A tribal mind-set cannot be reattained; we cannot turn back. It can only come as an echo, a hint in our armored individuated world that a bit of confusion as to ego boundaries can be an act of health, of homage and love, and can be a whisper of what it feels like to be swaddled in continuity. It is a lesson, amid our ever-expanding array of scientific labels, on the risks of overpathologizing. Most of all, it is a lesson that it would not be so bad, in fact it would even be a point of pride, if in the end, someone mistakes you for him.

For my father, Thomas Sapolsky, 1911–1994.

FURTHER READING

For an overview of the split-brain literature, see two books by Sperry's principal student and collaborator in the work: M. Gazzaniga, *The Bisected Brain*

(New York: Appleton-Century-Crofts, 1970); and M. Gazzaniga, *The Cognitive Neurosciences* (Cambridge, Mass.: MIT Press, 1995). For a summary of Sperry's view of the work, see R. Sperry, "Consciousness, Personal Identity and the Divided Brain," *Neuropsychologia* 22 (1984): 661. And for discussion of laterality of function in the brain, see N. Geschwind, "Cerebral Dominance in Biological Perspective," *Neuropsychologia* 22 (1984): 675.

For Jaynes's controversial ideas, see his 1977 book: J. Jaynes, *The Origin of Consciousness in the Breakdown of the Bicameral Mind* (Boston: Houghton Mifflin, 1977).

For a discussion of multiple personality and other "dissociative" disorders, see D. Spiegel, *Dissociation.* (Washington, D.C.: American Psychiatric Press, 1994). Dr. Spiegel was the psychiatrist paraphrased who contributed heavily to the new formulation concerning multiple personality disorder. This can be found in *Diagnostic and Statistical Manual of Mental Disorders,* 4th ed. (Washington, D.C.: American Psychiatric Association, 1994).

For Freud, see "Mourning and Melancholia," in *The Collected Papers,* vol. 4 (New York: Basic Books, 1959).

The Mann quote comes from *Joseph and His Brothers* (New York: Knopf, 1934).

Why You Feel Crummy
When You're Sick

Robert Longo, Pressure, *1982–83;*
courtesy Metro Pictures, New York

Recently, in preparation for a trip to the tropics, I received a dreadful number of inoculations. As I sat in the clinic after the shots, shifting from cheek to cheek in discomfort, the nurse explained that the vaccines—especially the one for typhoid—could make me feel a bit ill. And, sure enough, by nightfall I felt crummy.

It was a particularly unpleasant bout, because somehow it felt illegitimate. I wasn't really ill. I didn't have to worry that I had malaria or the flu or the plague. I knew what the cause was, and it wasn't serious. Vaccines, in essence, work because they fool the body into thinking it's just a little bit sick and had better defend itself. We're generally unaware of the mock battle being staged, but some vaccines, including typhoid shots, produce an unusually strong effect. So I couldn't really feel sorry for myself, and besides, I knew I'd be fine in the morning. In an odd way it was all form and no content—I felt sick without actually being sick.

An extraordinary array of illnesses make us feel crummy in this way. We want to sleep at all hours of the day. Our joints ache, and we feel cold and feverish. Sex loses its appeal. We lose interest in food; if the illness persists, we lose weight even if we force ourselves to eat. And we look like hell. Not long ago Benjamin Hart, a veterinarian at the University of California at Davis, listed sixty-odd common diseases in mammals that produce the same array of symptoms, despite their infecting different organs. Give a person influenza, which affects the respiratory system, give a cat infectious anemia, which affects its blood, give a sheep enterotoxemia, which affects its gut, and all will get achy and mopey and feel like putting on flannel pajamas.

Traditionally, such symptoms were not considered terribly interesting from a medical standpoint. Suppose you had the flu and complained to your doctor that you felt weak and your joints ached. I'd wager that the most common answer would be "Of course you feel weak. Of course your joints ache. You're sick." The symptoms are so nonspecific and ubiquitous that they seem almost not to count. But in recent years researchers have discovered a great deal about why we feel crummy when we have infectious diseases. These symptoms don't just happen; instead, your body works to bring them about—and, it seems, for very sound reasons.

At the center of this story is the immune system and the various white blood cells it employs to fight off disease. When a pathogen—an infectious bug such as a virus or bacterium— invades the body, it is the immune system that first sounds the alarm: the invader is promptly grabbed by a large scavenger cell called a macrophage. The macrophage in turn presents the bug to a helper T cell, which signals that the noxious foreigner is indeed worth getting excited about. The macrophage then sets off a chain of events culminating in the activation of killer T cells that attack the intruder. This cascade is referred to as cell-mediated immunity.

Meanwhile, a second form of defense, known as humoral immunity, is also set in motion: the helper T cells stimulate yet another type of white blood cell—B cells—to divide, differentiate, and ultimately produce antibodies to the intruder. These in turn will grab hold of the infectious organism and immobilize it.

Sorting out this process has kept immunologists off the streets for decades. What they've learned is that the immune response involves a variety of cell types scattered throughout the body. To communicate with far-flung members, the immune system uses cytokines, chemical messengers that travel in the bloodstream and lymph fluid. And that's where feeling crummy begins to come into the picture.

Among the best known of these messengers are the interferons, which activate a type of white blood cell that fights viruses and cancer, and the interleukins, which are central to the T cell cascade. The interleukin of greatest interest to us here is called IL-1 (for interleukin-1); its principal job is to carry the alarm message from the macrophage (where it is made) to the T cells. But we begin to feel crummy because that is not all that IL-1 can do; it can also influence the brain.

Most dramatically, this interleukin alters temperature regulation. For years it was known that, after infection, the immune system produced something that caused fever. No one knew what it was, and the putative messenger was called endogenous pyrogen, a term whose derivation should be familiar to pyromaniacs and Pyrex manufacturers among the readership. Not until the early 1980s was IL-1 identified as the pyrogen.

A part of our brain called the hypothalamus functions much like a thermostat. Normally it is set for 98.6 degrees. If body temperature drops below that, you shiver to generate heat, divert blood from the periphery of your body to vital organs, and pile on the blankets. Temperatures above 98.6 cause you to sweat and breathe faster to dissipate heat. What IL-1 does is cause the set point to shift upward. In other words, you begin to feel cold at 98.6, the various warming responses kick in, and a new equilibrium is reached at a higher temperature. You are now running a fever.

But that's not the only way IL-1 makes us feel crummy. A

few years ago two research groups (my own and a group in Europe) simultaneously reported that the chemical also causes the hypothalamus to release a substance called corticotropin-releasing factor, or CRF. This substance runs the body's hormonal response to stress by initiating a cascade of signals going from the hypothalamus to the pituitary gland, and from there to the adrenal glands, which prepare you for an emergency.

Suppose you walk into the supermarket and, unexpectedly, find yourself being chased by a rhinoceros intent on goring you. You will start secreting CRF within seconds—and for good reason. Corticotropin-releasing factor blocks energy storage, specifically inhibiting the process by which the body stores fat as triglycerides and sugar as glycogen. Instead, energy is diverted in a hurry to the muscles that are hurtling you down the produce aisle. At the same time, CRF dampens appetite, sexual drive, and reproductive processes. These effects are particularly logical; this would be a foolish time to waste energy by ovulating or planning lunch.

And it is much the same during an infection. Interleukin-1 triggers CRF release, and soon enough neither sex nor food seems particularly appetizing. Levels of sex hormones plummet, and if the illness lasts long enough, then ovulation and sperm production can be disrupted.

More crummy symptoms can be attributed to IL-1. It makes you sleepy—although no one is quite sure how. And it does something else particularly unpleasant.

There are nerve pathways coming from various outposts in your body—from the surface of your skin to deep within your muscles and tendons—that carry pain messages to your spinal cord; these messages are then relayed to the brain, which interprets them as painful. Obviously, stepping on a tack is a very painful stimulus: the pathway originating in your toe is going to be activated, and your brain will register pain in an instant. But far more subtle stimuli will not cross a pathway's activation threshold. They will not be perceived as pain unless the threshold is lowered. And that's exactly what IL-1 does; it makes the neurons along the pathway more excitable, inclined to react to things they would normally ignore. Suddenly, your joints hurt, old injuries ache again, your eyeballs throb.

Altogether, that's quite an impressive array of effects for a chemical that was once thought of only as an immune system messenger. And it's this very quality that makes interleukins unpleasant to use as treatments for illness. Potentially these chemicals could help you fight disease by stimulating your immune system. But such a treatment is guaranteed to make you feel crummy. Cancer patients who have been given IL-2, a close relative of IL-1, have felt stupendously sick.

Biologists have learned a bit about just how IL-1 triggers these crummy symptoms. The chemical binds to receptors on the surfaces of those neurons that play a role in perceiving pain, regulating temperature, and releasing CRF. That, in turn, switches on the synthesis of prostaglandins, compounds that act as signals inside the cell to change temperature set points and sensitivity to pain messages. So potentially, you could decrease a lot of the symptoms of feeling crummy if you took a drug that blocked that prostaglandin synthesis. Which is just what is accomplished by aspirin.

One major aspect of the sick syndrome still needs to be explained: cachexia, or the wasting away of body weight. This is an obvious thing to happen during a sustained illness, given what happens to your appetite. But cachexia defines something more. During a chronic illness you lose weight even faster than can be explained by decreased eating. Your body has trouble storing energy.

Corticotropin-releasing factor appears to account for some of that. Recall your desperate flight from the rhino in the super-market: during such an emergency you need energy for your

muscles at that instant, not later, and CRF, as we saw, indirectly blocks energy storage. But CRF also causes stored energy
to be released, returning fuel in the form of fat and sugar to the
bloodstream. Thanks to IL-1 and hormones such as CRF, the
same thing happens during an infection; if it goes on long
enough, you lose your fat stores and begin to waste away.

But cachexia is also caused much more directly. When a
noxious agent is first spotted in the body, macrophages secrete
a second cytokine along with IL-1. In addition to its role as an
immune system messenger, this cytokine blocks the ability of
fat cells to store fat. Reasonably enough, this chemical has
been dubbed cachectin.

Looking at such a wide range of effects, it seems difficult to
argue that feeling crummy is something that just happens for
no good reason. Clearly, the body works very actively to bring
those symptoms about. Various brain regions have evolved
receptors for IL-1, and fat cells have developed specific mechanisms for responding to cachectin. Your body is working in a
very complex and specific way to make sure you feel awful and
waste away when you are sick. What could be the use of that?

Some of the symptoms make sense. A showdown with a
virulent pathogen can require as much energy as a showdown
with a rhino. It is no small task to trigger the massive prolifer-
ation of immune cells needed to mount a defense against
infection: cells must divide and migrate at a tremendous rate;
cytokines and antibodies must be hurriedly synthesized and
secreted. All this immune activity does not come cheap; it
requires a great deal of instant energy. Thus it is quite helpful
for CRF and cachectin to block energy storage and keep fuel
readily available. It is also quite logical to inhibit reproduction
if you're sick, since producing offspring is one of the most
expensive things you can ever attempt with your body, especially if you are female—you're better off using the energy to
fight the disease. If the disease is going to be around for a

while, it is a particularly inauspicious time to get pregnant anyway.

By the same logic, it is probably a good idea to sleep more during an illness, because sleeping conserves energy. And maybe your joints hurt during a fever to make you conserve energy also—I'm not inclined to run around a lot when I ache all over. But it's unclear why you should lose your appetite, especially since many features of feeling crummy seem geared toward meeting the need for energy. I'm not particularly convinced by the hypotheses I've seen advanced. One, for example, holds that appetite loss helps animal species that are preyed upon by carnivores. Otherwise these animals would be out foraging for food when they are obviously under the weather, inviting predators to attack them. The trouble with this is that card-carrying carnivores also lose their appetite when they're sick.

The most dramatic feature of feeling crummy, of course, is fever. The energy mobilized during an infection goes to fueling not only the immune system but the shivering muscles as well. During a malarial fever, for example, metabolism increases by nearly 50 percent, and much of this energy expenditure goes toward generating heat. An investment on this scale certainly suggests fever is doing something useful for you. And indeed a number of studies support the idea.

The easiest way to demonstrate the benefits of feverishness would be to infect laboratory animals with something that typically causes a fever and see how they fare when the fever is blocked. You might, for example, administer an antipyretic drug such as aspirin and then monitor what happens to the animals' immune response, antibody concentrations in the bloodstream, and survivorship. If those measures are worsened, you can conclude that the fever normally helps fight the infection. However, aspirin does a lot of things besides decrease fever (for example, it decreases joint pain), and its antifever effects may not be the cause of any changes you observe. To get around the problem, Matthew Kluger, a physiologist at the University of Michigan Medical School, did one of the most supremely clever studies that I have ever heard of.

Kluger studied lizards—which are, of course, cold-blooded—placing them in a terrarium that was hot at one end and cold at the other, and had a smooth temperature gradient in between. When healthy, the lizards settled down at the point in the gradient where their body temperature would stabilize at 98.6 degrees. But when Kluger infected them with a bacterium, they chose to become feverish—they moved up the gradient to a point where their temperature would rise a few degrees. When Kluger prevented the infected lizards from moving to the hotter end of the terrarium, they were less likely to survive the infection. Running a fever, he concluded, helped.

The reasons for this are at least twofold. First, the immune system works better when you are running a fever. Studies show that T cells multiply more readily and antibody production is stepped up. Second, a fever puts many pathogens at a disadvantage. A wide variety of viruses and bacteria multiply most efficiently at temperatures below 98.6 degrees. But as a fever is induced, their doubling time slows. In some cases the pathogens stop dividing entirely.

Such studies suggest that fever and the host of other changes we suffer are adaptive mechanisms for helping us through the challenge of infection. It's not a perfect plan; not all bugs, for example, are inhibited by heat, and too high a fever will damage you along with the infectious invader. But as a general strategy it seems to work.

These observations suggest that fever-reducing drugs such as aspirin may not always be such a swell idea. It would be ironic if future research concludes that the best thing to do during an illness is simply to endure feeling crummy. But perhaps you'll at least draw some comfort in knowing that, as a result of your stoicism, the pathogen is feeling even crummier.

EPILOGUE

Since this piece was published, a trend has emerged that will horrify the scientifically fainthearted. About a decade ago, it was still quite novel to think that immune messengers could also influence brain function, behavior, hormone release, fat metabolism, and other nonimmune aspects of physiology. While this was new and puzzling, it was at least relatively clear-cut, in that IL-1 seemed to account for the fever, CRF release, pain, and sleepiness, while cachectin had dibs on cachexia. Immunologists have been nearly overwhelmed by the discovery of dozens of different cytokines (for example, the interleukins are up to at least the teens in numbers). Naturally, in the last few years, it has turned out that a lot of these cytokines also participate in the regulation of fever, CRF release, pain sensitivity, etc. Some of these cytokines synergize with IL-1 or cachectin, some work independently, and it's all very confusing. While it is exhausting to try to absorb all these new individual facts, the general picture remains the same—the generic symptoms of feeling crummy don't just happen to occur offhandedly, but instead arise for adaptive reasons thanks to some very explicit and clever tricks on the part of the immune system.

FURTHER READING

Benjamin Hart's work on how a wide variety of diseases induces the same generalized crummy symptoms is summarized in B. Hart, "Biological Basis of the Behavior of Sick Animals," *Neuroscience and Biobehavioral Reviews* 12 (1988): 123.

For a general introduction for the nonspecialist to the workings of the immune system, see chapter 8 in R. Sapolsky, *Why Zebras Don't Get Ulcers: A Guide to Stress, Stress-Related Diseases, and Coping.* (New York: W. H. Freeman, 1994). Chapter 2 also gives a general overview of the workings of

CRF and glucocorticoids. For a more detailed overview of the former, see A. Dunn, "Physiological and Behavioral Responses to Corticotropin-Releasing Factor Administration: Is CRF a Mediator of Anxiety or Stress Response?" *Brain Research Reviews* 15 (1990): 71.

The finding that IL-1 releases CRF from the brain was first reported in R. Sapolsky, C. Rivier, G. Yamamoto, P. Plotsky, and W. Vale, "Interleukin-1 Stimulates the Secretion of Hypothalamic Corticotropin-Releasing Factor," *Science* 238 (1987): 522; and F. Berkenbosch, J. van Oeers, A. del Rey, F. Tilders, and H. Besedovsky, "Corticotropin-Releasing Factor–Producing Neurons in the Rat Activated by Interleukin-1," *Science* 238 (1987): 524.

The clever fever study is described in M. Kluger, "The Evolution and Adaptive Value of Fever," *American Scientist* 66 (1978): 38.

The roles of IL-1 and related cytokines as pyrogens, somnogens, and sensitizers to pain are reviewed in a number of technical papers: R. Bellomo, "The Cytokine Network in the Critically Ill," *Anaesthesia and Intensive Care* 20 (1992): 288; F. Obal, J. Fang, L. Payne, and J. Krueger, "Growth-Hormone-Releasing Hormone Mediates the Sleep-Promoting Activity of Interleukin-1 in Rats," *Neuroendocrinology* 61 (1995): 559; J. Krueger and J. Majde, "Microbial Products and Cytokines in Sleep and Fever Regulation," *Critical Reviews in Immunology* 14 (1994): 355.

The ability of cachectin (also known as tumor necrosis factor) to induce cachexia is reviewed by the discoverers of the phenomenon: B. Beutler and A. Cerami, "Cachectin and Tumour Necrosis Factor as Two Sides of the Same Biological Coin," *Nature* 320 (1986): 584. See also: A. Cerami, "Inflammatory Cytokines," *Clinical Immunology and Immunopathology* 62 (1992): S3. This is a case of a pattern that frequently pops up in this business. For years, scientists interested in how the immune system can kill some types of tumors studied a compound critical to that, called tumor necrosis factor. Meanwhile, in the next county, other scientists, interested in how the immune system could regulate metabolism and fat storage, were busy with cachectin. Eventually, both groups were able to determine the structures of their compounds and, surprise . . . tumor necrosis factor and cachectin turned out to be one and the same. (This, predictably, led to some tense negotiations as to which name the compound should be called. In a Solomonic solution, it is now called by both names, thereby avoiding needless bloodshed among immunologists.) More and more, immune and endocrine messengers turn out to be versatile, having multiple functions.

Circling the Blanket for God

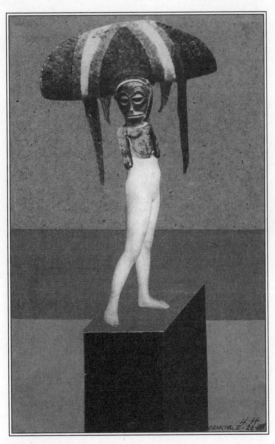

Hannah Höch, Denkmal II: Eitelkeit *(Monument II: Vanity), 1926;*
copyright 1996 Artists Rights Society (ARS) NY/VG Bild-Kunst, Bonn

A warning: If you've gotten this far in this collection, I can now number you as one of the few faithful readers of this stuff who is not either a close relative or a student trying to get a good grade in my class. Thus, you'll have to trust me on this and promise to do a few things. First off, if you think you would be offended by an analysis of religious belief in a medical or psychological context, don't read this. I'm not kidding. However, if you do start reading this and get offended, please, please read this through to the end and see if it still seems as disturbing, or at least disturbing in a different way than you had originally thought.

I. Hearing Voices at the Right Time

History has not been kind to Paul Radin, and I begin by trying to rectify this a bit. The accomplishments of this obscure anthropologist, trained in the school of Franz Boas at the beginning of this century, are at least twofold. First, by dint of having been born to a rabbinic family in Lodz, Poland, and by then devoting his professional career to the anthrolinguistics of Native American tribes of the plains, Radin was probably the only human to have ever walked this planet who could have produced a reliable Yiddish-Sioux phrase book. He never did so. In fact, there is no evidence that he even so much as gave a moment's thought to the proper Sioux translation of *schlemiel* or *mensch,* on the one hand, or the most appropriate Yiddish equivalent of *tepee.* Nonetheless, the mere fact that he could should entitle Radin to membership in the tentatively emerging pantheon of multiculturalism.

Radin's second accomplishment was a bit weightier. In 1936, he wrote a book containing an idea that, if taken to its logical conclusion, should undo some of the most cherished underpinnings of Judeo-Christian thought.

Radin's speculations indirectly cast light on the question of why schizophrenia exists. This is not necessarily an obvious question to ask. Schizophrenia is one of the most catastrophic ways in which the mind can go awry. It is worth reviewing the features of the disease in order to appreciate Radin's insight and its potential consequences. This is also useful to help counteract an odd misuse of the word "schizophrenia" by the media and by nonexperts. The erroneous view is of schizophrenia as a disease of sudden, unpredictable lurches between emotional extremes—"God, what a schizophrenic day I'm having. First my car wouldn't start this morning and I got to work really late. But then I landed this account I've been angling for, for months. Then, at lunch I had this really upsetting argument with a friend. What a schizy day."

In contrast to this incorrect picture, schizophrenia is a disease of disordered thought. The marvel of it is that the cognitive tumult is not just random, but has surprisingly consistent patterns from one sufferer to the next.

Above all else, schizophrenics show "loose associations." Most of us might relate an incident in a way where there is an obvious logical progression from one step to the next. Schizophrenics, in contrast, produce a storm of non sequiturs or make leaps that, while having a certain strained connectiveness, are not the ones anyone else would make: "So I guess I've been a real disappointment to my parents. After all they did for me to get a good education, you know what I've wound up doing? I'm a caddie. A caddie. And no one wants caddies these days. Everyone wants the Japanese cars."

Schizophrenics also have problems with levels of abstraction. Most of us can hear a story and intuit readily whether it is meant to be a factual, literal relating of events, or a parable, meant to be taken symbolically. We intuit whether the point of a particular story is the trees or the forest. Schizophrenics lack that intuition, and their bias is to perceive things on the far more literal, detailed level. "Concreteness" is the term for it and it manifests itself endlessly:

THERAPIST: Hello, Mr. Smith, what's on your mind today?
PATIENT: My skull.

Or:

THERAPIST: Tell me, what do apples, bananas, and oranges have in common?
PATIENT: They all are multisyllabic words.
THERAPIST: Anything else?
PATIENT: They all contain letters that form closed loops.

For schizophrenics, it's not a matter of trees and forests. Instead, it's habitually seeing only the bark.

Schizophrenics are also prone toward delusions, inserting themselves into situations, making claims that cannot be so, and seemingly believing them. ("Heard of the Great Wall of China? Yes? My idea. The emperor came to me at night with a map and I said, 'Here's where it goes.' ") Related to this are the hallucinations, predominantly auditory, that are defining features of the disease. And related to this is the tendency in many schizophrenics toward a florid, permeating paranoia. ("What do apples, bananas, and oranges have in common?" "They're all wired for sound.")

Add to that some of the other features of the disease less

related to thought—an inappropriate flattening of emotions, an aching social isolation, a tendency toward horrifying acts of self-mutilation and suicidalism—and you have one of the great medical tragedies to befall a person.

One might ask why should schizophrenia exist in culture after culture on this planet, as it does. The disorder has a genetic component (and by this I mean the modern view of human behavioral genetics—not that there is *a* gene that inevitably *causes* schizophrenia, but that there are gene*s* that make the individual more *sensitive* to schizophrenogenic triggers in the environment). The disorder runs in families. Studies of adopted individuals showed that the trait is more common among the genetic, biological pedigree, rather than through the nongenetic, adoptive one. And currently, molecular biologists search for the actual genes, the precise DNA sequences, relevant to the disease.

Thus, if there is a genetic component to the disease, no matter how small, when one asks, Why should schizophrenia exist? one is actually asking, Why should the gene(s) involved in schizophrenia have evolved and been maintained by natural selection in the human gene pool?

Evolution, we were all taught in ninth-grade biology, is the process by which genetic traits are passed on over time and become more common only if they are advantageous. Once, long ago, when giraffes had the short necks of most mammals, one had a somewhat longer neck for a genetically based reason. And that giraffe could reach leaves higher up in the tree, was fruitful, and multiplied . . . and now it is that giraffe's long-necked descendants who stroll around the savannah. Evolution, we were taught, works something along those lines. So what's the evolutionary advantage to schizophrenic genes? Schizophrenics are "unfit" in strict evolutionary terms, which is to say that they have a lower reproductive rate, and thus pass on fewer copies of their genes, than do

healthy individuals. Then why is this disorder being maintained in the population?

The answer probably lies in a facet of genetics that we all
also learned in ninth grade. Sometimes, genetic traits can
come in different degrees of severity, and while the full-blown
version of such a trait may be disadvantageous, the moder-
ately "penetrant" version that occurs in a relative may carry a
big advantage. And if the magnitude and frequency of the
advantageous form outweighs the deleterious, the trait will be
selected for in the population. The classic example is sickle-
cell anemia affecting people of African descent. In its extreme
form, it is a lethal hematologic disorder, while in its less penetrant form, it protects against malaria. There is evidence for a
similar pattern in Tay-Sachs disease among Ashkenazi Jews.
In its full-blown form, it is a fatal neurological disaster. In its
milder form, some data suggest, it confers protection against
tuberculosis. There's also the suggestion that the gene for cys-
tic fibrosis protects against cholera.

Schizophrenia probably shows this pattern. What would be
the mild, advantageous version of schizophrenia? As was mentioned in the first piece in this volume, this form is now called
"schizotypal" personality disorder. Schizotypals are not the
dysfunctional isolates that schizophrenics are. They just tend
toward solitary hobbies and professions, are uncomfortable in
social situations involving unfamiliar people, are aloof, with few
close friends. They are the fire tower rangers, the lighthouse
keepers, the film projectionists alone each night in their booths.
Moreover, schizotypals are not floridly delusional, hallucinating
like schizophrenics. Their disordered thoughts and behaviors
are far subtler, in that they have a tendency toward what is
termed the "metamagical." They may have an extremely
strong interest in science fiction and fantasy, or in some New
Age paranormal belief such as ESP or levitation. They often
report odd perceptual experiences—sensing spirits in a room,

seeing illusions. Or, as a foreshadowing of the tumult to come, they may have a profound faith in very literal, very concrete interpretations of religious dogma—it really is possible for Jesus to have walked on water, the Patriarchs truly lived for nine hundred years, the creation of the world in seven days is reporting of precise fact rather than a parable. Schizotypal personality disorder was first recognized in precisely the population where you would expect it—among close relatives of schizophrenics, among those individuals who share a certain percentage of genes with schizophrenics.

It is important to recognize that none of these traits count as all-out mentally ill in the conventional sense used by most in society, in which "crazy" is most usually akin to the shattering psychosis of schizophrenia. Perfectly respectable businessmen may sneak off to their Star Trek conventions, an erstwhile actress may publish a best-selling account of her previous lives, a First Lady may consult astrologers and still be taken seriously in the fashion pages.

Who are the schizotypals? Not the lone schizotypal operating the film projector in a movie theater, sensing the presence of Elvis in the room. I mean who were the schizotypals in preindustrial societies, throughout 99 percent of our human history? Here is the key. In 1936, Radin was the first to advance the idea that many shamans, witch doctors, and medicine men (and women) are "half-crazy."

It fits. The shamans, the forbidding, charismatic religious leaders in tribal life, the ones who sit and converse with the dead ancestors, who have solitary sojourns in the desert, whose huts sit separate from everyone else's, who spend the night transformed into wolves or bears or hyenas, the ones who lead the trance dances and talk in tongues and bring word of the wishes of the gods.

It was Radin who first focused attention on the idea that, in Western societies, shamans would be viewed as a bit psychiatrically suspect. "Throughout the world of primitive man some form of emotional instability and well-marked sensitivity has always been predicated as the essential trait of the medicine-man and shaman." He branded the shaman with a label that, based on my perusal of the psychiatric texts of his time, I suspect was invented—"neurotic-epileptogenic." I think this terminology reflected an odd, tragic chapter in the psychiatry of his era, in which epilepsy was considered a psychiatric rather than neurologic disorder, and where the pre-seizure auras of the epileptic were often grouped with the auras and hallucinations of the schizophrenic.

Importantly, Radin focused on the notion not only of the shaman as "half-crazy," but of his instabilities as the creative seeds of future religious convention. "[The shaman displays his possession by a spirit] by publicly reenacting his specific personal experience, that of a man suffering from a particular mental affliction. His projections, his hallucinations, his journey through space and time, thus became a dramatic ritual and served as the prototype for all future concepts of the religious road of perfection." If you aspire to shape supernatural belief in your society for generations to come, some half-mad inspiration would help.

Radin's ideas were soon echoed by others. Erwin Ackerknecht, a physician and naturalist, discussed the shaman as the "healed madman." Jules Silverman, a psychotherapist, weighed in with a similar view and explicitly compared the symptoms of the schizophrenic with the traits of the shaman. In a rather pungent statement, the psychoanalyst George Devereux wrote: "[Primitive] religion and in general 'quaint' primitive areas are organized schizophrenia." By the 1960s, the anthropologist Richard Shweder was even conducting empirical studies in the field, demonstrating that shamans have different cognitive styles than other members of their societies.

Probably the most insightful elaborations on Radin's

notion came in 1940 from Alfred Kroeber, one of the biggest guns in anthropology. "In some cultures one of the most respected and rewarded statuses known to the society is acquired only by experience of a condition which in our culture we could not label anything else than psychotic."

Kroeber emphasized a number of points. While Radin had focused on the generative and creative potential of the psychiatric afflictions of the shaman, Kroeber was more impressed with the fact that these afflictions are nonetheless framed within a preexisting cultural framework. You do not merely report on your time as a wolf or burst out in glossolalic babbling in order to get licensed as a witch doctor. Rather, there are rules as to how shamans go about being psychiatrically unruly. Kroeber discussed how in endless cultures, young individuals are recognized by the established shamans as showing the first signs of the hallucinations and psychoses and are thereafter trained in the particular patterns of shamanistic experience in that culture. Thus, the schizotypal traits are channeled and standardized.

As a second point, Kroeber dwelled on the rewards of such psychosis. "To us a person that hears the dead speak or proclaims that he sometimes turns into a bear is socially abnormal, at best useless, and likely to be a burden or menace." Shamans, instead, are anything but that. They are powerful, honored, feared, sought after. Shamanism is a highly rewarded state. It is true that in many traditional societies, such shamanism is associated with a reclusiveness, a social withdrawal often linked to celibacy. But in at least as many cultures, shamans and their kin are rewarded not just materially and with respect, but reproductively as well. Sure, you don't want everyone to be a shaman. This was stated well by a Winnebago tribesman who turned to Radin during a particularly frothy shamanistic ceremony and said, "Well, it's good that some of us are that way some of the time, but it would be disastrous if all of us were that way all of the time." However, having the occasional shaman is highly valued by traditional societies, and these shamans are typically powerful, respected members. Insofar as shamanism and its attendant hallucinations, metamagical thought, and psychiatric instability reflect some genetic component, these are not unfit traits being winnowed from the gene pool.

Kroeber encapsulated his idea of the acceptance and rewards of the traits of the shaman in the title of his essay, "Psychosis or Social Sanction." That final word gave me a poignant twinge when considering its similar derivation as "sanctuary," and the rarity with which the most modern of psychiatry can provide that for our mentally ill.

In the years since Radin and Kroeber, most anthropologists have distanced themselves from the strongest versions of their ideas. There is, in fact, next to no evidence in any culture that shamanism is "*always* predicated" on psychiatric instability, that such status is "acquired *only* by" psychosis. In most traditional cultures, the majority of shamans appear to be quite mentally healthy and are often orphans who had been raised by shamans as surrogate parents. It is probably these average-joe shamans, punching in the ol' nine to five clock, who mostly ply their trade through the classic shamanistic image of "tricksterism"—sleight of hand, ventriloquism, a certain degree of manipulative showmanship. There is little evidence that they are wired to have different cognitive styles from everyone else, and there are virtually no anthropologists who are trying to find that out. The main point of Radin's and Kroeber's ideas is not that all shamans are "half-crazy," but that if you are half-crazy in the right way, shamanism constitutes a uniquely protected and rewarding refuge.

Kroeber emphasized an additional and critical point in his writing on the subject. He made clear that even the most floridly addled of shamans are typically not what we would now classify as schizophrenic, at the extremes of psychotic dysfunction. He noted how even traditional societies have little use for someone who perpetually believes himself to be a tree and acts as such. As a by-product of some of my own research in East Africa, I have had the occasion to observe that traditional societies there are no less unnerved by and intolerant of schizophrenics than ours is. But these are not shamans, whose psychoses are more controlled. Shamans hear voices during times of crisis rather than all the time, or bellow in tongues during ceremonies rather than at the critical silent moment during a hunt. "In general, the psychopathologies that get rewarded among primitives are only the mild and transient ones," Kroeber wrote. Years before the invention of the term "schizotypal," Kroeber had linked shamans to that classification rather than to schizophrenia.

II. The Evolution of National Holidays in Western Societies

Thus, the psychopathology theory of shamanism of Radin and others suggests that mild forms of schizophrenia lead one toward the highly adaptive role of being a shaman in traditional society. Stated in modern terms, schizotypal personality disorder and schizophrenia are on a genetic continuum, and the selective advantages of the former are sufficient to maintain the broad genetic trait, including the maladaptive schizophrenic variant, in the human gene pool. Or, to put it more simply, having the occasional cousin who is schizophrenic is the evolutionarily tolerable price of having a ready supply of schizotypals.

These musings were presented with a subtext whose error forms the heart of this piece. There is a trend in some circles to use contemporary multicultural standards and catchwords as a means to wield a pretty censorious tone about our intellectual forebears (read: dead white males). This is often politically

Eugène Delacroix, Jacob Wrestling with the Angel *(detail)*, 1853–63;
St. Suplice, Paris, Giraudon/Art Resource, New York

correct giddiness. Nevertheless, one cannot help but recoil at
the foundations of racism that run through the history of
social anthropology.

Many of the thinkers discussed took pains to note how
uniquely the psychopathology of the shaman is an issue of the
folks with the bones in their noses. As is clear from some of the
quotes above, "primitive" runs through all these considerations.
Radin imposed a pseudomathematics on this judgmentalism,
claiming that there was a quantitative correlation between
how "simple" a society is and the extent to which the shaman
class is dominated by the half-crazy. My favorite version of this
was in Kroeber's work, which discussed the psychotic shaman
as a feature of "low cultures," in contrast to the higher cultures
freed of them. Toward the end of his essay, he went off on a
humane, impassioned lament about the racism of prior anthropology, its ethnocentrism in imposing Western standards of

progress. Having gotten that off his chest, Kroeber then clearly links "progress" of a society to the rejection of shamanistic nonsense and its attendant confusion of reality with psychotic subjectivity. "I cannot rid myself of the conviction that there does exist a real relation between on the one hand certain symptoms justly reckoned psychopathological . . . and the degree of cultural advancement or progress."

Some contemporaries of these individuals attacked this racism. In 1946, Ackerknecht deplored the statement that "primitive" religion is nothing but organized schizophrenia, claiming that only the gigantic vanity of our own culture could give rise to that idea. It seems to me that Radin and Kroeber were wrong in distinguishing between "primitive" theology and our own. But not wrong in the way suggested by Ackerknecht, who in effect said, How dare you imply that the theology of those cultures is shot through with a psychiatrically suspect confusion of subjectivity with reality? Their theology sure as hell is shot through with such confusion. The main point is, so is that of Western societies. As seems intrinsic to the whole process of religious belief.

The most provocative implication of the theory of shamanistic psychopathology is that it is not just about the *Clan of the Cave Bear* or the folks with no clothes in the *National Geographic*s. Ackerknecht noted the demise of supernaturalism in Western societies. "Our culture is unique in its consequent outlawing of the irrational." The reports of the death of such subjective confusion are greatly exaggerated. A few statistics that will perhaps shock the sheltered readership to as great an extent as they did me: In a recent Gallup poll, 25 percent of Americans fessed up to believing in ghosts, while 36 percent admitted to believing in mental telepathy, 47 percent in UFOs, and 49 percent in extrasensory perception. A majority of the citizenry of our fair land claimed to believe in the devil and 49 percent believed that it is possible for someone to be possessed by said devil. Mind you, this was a poll of American *adults* (i.e., individuals generally allowed to vote, operate motor vehicles, or serve on juries). The mind, at least that of this reporter, reels.

Our daily lives may not be as permeated with superstition and supernatural belief as are those of many in traditional societies, but our society on the threshold of the next millennium is nevertheless splotched with irrational subjectivity. Some variants of such irrationality have been concocted out of air in recent times. This is the silly world of crystals and channeling and healing pyramids, the fast-food spiritualism of nouveau medieval peasants in hot tubs. But some of the beliefs whose embrace cause vast numbers of us to abandon rationality come with pedigrees stretching back millennia.

The notion of the psychopathology of the shaman works just as readily in understanding the roots of major Western religions as well. This is the first place where I must tread carefully, as many readers will no doubt be offended by these ideas, despite my warnings at the beginning of this piece. Put succinctly, it is not usually considered to be a sign of robust mental health to hear voices in burning bushes. Or to report that you've spent the night wrestling with an angel, or that someone who has died has risen and conversed with you. Who originated those beliefs? Who first reported the number of days it took for the world to be made? Who first emerged with the fact that there was once an Edenic garden and that snakes bearing apples are up to no good? Who announced that virgins can give birth? These weren't the products of some committee after careful market research. Each had to have come from the metamagical subjectivity of some formative and formidable visionary, whose identity is usually lost to us forever but whose fevers took hold.

So long as one bypasses the idea that there is occurring literal reportage of divinely dictated events, there appears to be a rich vein of schizotypal behavior and belief in the cornerstones of theology of major religions. And the rewards of it are no less than in non-Western culture. Oh, sure, one can overdo it, and our history is darkly stained with abortive religious movements inspired by messianic crackpots. But it appears to be a continuum: too much, and you are arguably in the realm of a Jim Jones, David Koresh, or Charles Manson, all of whom were able to lead followers into a maelstrom of paranoid delusion. In the cases of Jones and Koresh, one can only do armchair forensic psychiatry to try to guess their afflictions, but Manson, alive and well, is a diagnosed schizophrenic. However, if you get the metamagical thoughts and behaviors to the right extent and at the right time and place, then people might just get the day off from work on your birthday for a long time to come.

III. Cleanliness Really Is Next to Godliness

Thus, it seems to me that plausible links can be made among schizotypal behavior, metamagical thought, and the founding of certain religious beliefs in both non-Western and Western societies. This is typically big-picture theology—how the world begins and ends, what happens after you die, how many gods there are and how they get along with each other. These are steel girders, the superstructure of religious belief.

But religion is not just a handful of precepts that form the dominating core of one's particular beliefs—the Holy Trinity, or "I am that I am," or "There is but one God and he is Allah." Religion is not just a foundation of thought and faith. Nor is it just a set of moral imperatives, or a set of cultural values to be shared with a community. In its traditional, orthodox incarnations, it is also a collection of small habits, behaviors, and prohibitions, a myriad of everyday activities and sayings. "Religion is meant to be bread for daily use, not cake for special occasions," said Henry Ward Beecher. If the devil is in the details, then so is God, and for the average practitioner, religion is in the rituals and rules of quotidian life. It is here where I believe there is

Mark Tansey, Robbe-Grillet Cleansing Every Object in Sight, *1981;*
The Museum of Modern Art, New York, gift of Mr. and Mrs. Warren Brandt;
courtesy Curt Marcus Gallery, New York

a second critical link between a neuropsychiatric disorder and features of religious belief and behavior.

During periods of stress or of anxiety in our lives, most of us will find ourselves having certain small behaviors or thoughts that we cannot suppress, little rituals that, repeated once, are benign, but repeated at length, become irritants. Perhaps we are unable to stop from counting the number of steps we climb on a staircase or cannot stop some profoundly insipid and distracting commercial jingle from running through our heads. There are times when, to our visceral horror, we are unable to prevent ourselves from thinking something unthinkable— What if the person who is most important to me on earth suffered a horrible accident? What if that person were killed? You swim in the details of being notified by the police, of being accompanied to the morgue, of the inconceivable sight of a body—all the while thinking, Stop torturing yourself! Why are you thinking this? Stop it! and yet the thoughts leak on. We may not be able to begin working on a difficult and anxiety-provoking report until the pens and erasers at the desk have been arranged properly. Or when mailing off a critical application, we might check the slot on the mailbox a second or third time to assure ourselves that the letter has gone in and maybe even look behind and underneath the mailbox. Irritated, we may try to prevent ourselves from doing these things, but often, we can't help it. We check, just to make sure, just to make sure.

We are all subject to many of these small rituals. Here's an unexpected one that I have found to be remarkably widespread in children of late-twentieth-century American culture. Consider the following highly technical diagrams:

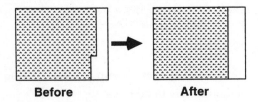

Before   **After**

The schematic on the left represents a tray of Kellogg's Rice Krispies marshmallow treat; the stippled pattern is the uneaten part. The schematic on the right represents what the Rice Krispies Treat looks like if you leave a person armed with a kitchen knife alone with it for a few minutes. *Most of us have the irrepressible urge to cut away and eat a sliver of the stuff so that it straightens out the edge.*

I have actually surveyed generations of my students as to why they carry out this behavior. Do they straighten out the edge so that their subsequent nibbling is less readily detected and somehow doesn't count as much as cheating on a diet? Rarely. Do they do their cutting so as to decrease the exposed length of edge, thereby preserving the freshness of the delicate slabs? Never. It's the same answer as I would give—there's just an itch, a sense of incompleteness, that makes it impossible to leave it alone. It's just not right to have an uneven edge of Rice Krispies Treat. The eating is almost an afterthought.

This is benign nonsense, as are those occasional, uncontrolled repetitions of thoughts or behaviors for most of us. They are generally thought to become more common in most of us at times of anxiety or stress because they provide a comforting sense of structure, at least about some little things, at a time when a Big Thing seems as structured and reliable as quicksand. But for a subset of us, these rituals of thought and behavior become a crippling psychiatric disorder.

It is called obsessive-compulsive disorder (OCD) and, as was noted in "How Big Is Yours?" it has moved from the backwaters of psychiatry to become almost a diagnostic fad. When thinking about the minor compulsive rituals that bother most of us now and then, it is hard to imagine the extent to which a life can be destroyed by these patterns writ large. Yet this is precisely what happens with the OCD sufferer. It is the severity of their bewildering symptoms that is most striking.

First and foremost, OCD patients must clean themselves. By this, I do not mean that they are fastidious, hygiene-conscious individuals. I mean that they wash their hands four, six hours a day, to the point where they cannot hold down a job for their time spent washing. They take a shower, methodically scrubbing with an array of carefully researched antiseptic soaps, each used in a proper sequence, applied in a particular manner to its designated part of the body, allowed to cleanse for a predetermined number of seconds before being washed off in water whose temperature has been picked for its antibacterial attributes. Finally, purified and fortified, the individual dresses and invariably, while leaving the room, brushes an elbow against the doorway. Who knows who else might have brushed against that spot, what germs they carried? Soiled, the person must bathe all over again.

As another frequent compulsion, the OCD individual checks. Vacations are never taken because of the endless examination of the window locks, the burner on the stove, the latch on the back door, before being able to leave; homework assignments or reports are never submitted and applications are never mailed off because they must always be checked once again for typos—there can never be enough certainty. The obsessive-compulsive will drive fifty miles out of his way, convinced that he has hit someone on the highway, that a body has flown off onto the divider, that he has unconscionably left the scene of an accident. He must go back and check and, finally convinced that there is no body and no heinous crime, he will finally drive on . . . only to become anxiously convinced that the body actually hurtled off in the other direction into the unexamined bushes on the shoulder. And thus he turns around again.

There are also compulsions of entering and leaving a place—a doorway can become an almost magically difficult threshold to cross until the individual touches the doorjamb a set number of times with the palm of her hand. There are rituals of symmetry and arranging of things—a meal cannot be begun until the utensils are perfectly, unattainably parallel. Intercalated in these rituals are ones of counting—light switches that must be turned on and off a set number of times, stairs that must be counted. In some more educated OCD patients, such numerology takes on arcane complexity. The OCD individual may find herself walking from her car to the entrance of the office building in which she is to interview for a job, and might impose a rule—she can only enter the building if she has walked an even number of steps, or a number of steps equal to a prime number, or to a prime number squared, or perhaps where the ratio of the number of steps in which her briefcase is carried in the right hand versus the left hand must conform to the Fibonacci ratio. Or else, she must start all over again . . . until she's late for another job interview. The task of protecting one's mother's back by not stepping on cracks becomes trivial in comparison.

These are the compulsions, the uncontrollable *acts*. Along with this comes uncontrollable obsessive *thoughts*. It is bad enough that such obsessions are often banal drivel of nonsense words or songs—it is dizzying to contemplate an adult, sentient mind obsessively consumed with the question, "Who put the bomp in the bomp-shoo-bomp-shoo-bomp?" But the obsessions are often darker. Perhaps the OCD individual cannot stop the thought of some tragedy befalling a loved one, of inadvertently being responsible for that tragedy, of intentionally doing the awful act. The person cannot stop the thought of a violent or perverse act, hours each day filled with the unstoppable details and the recriminations for thinking such filth. Or the individual may be overwrought with obsessive thoughts of guilt, a constant need to confess something.

Perhaps hardest to imagine for the outsider is the suffering of these people. In Tourette's syndrome, another disorder receiving a great deal of attention, individuals have a variety of emotive tics. The person curses uncontrollably, makes animal sounds, will briefly make a bizarre and strained facial expression. In Huntington's chorea, the patient might have uncontrollable swinging of the arms. But these stigmata are superimposed on the person. The Tourette's patient does not really wish to bark like a dog in some instant. It merely happens, an uncontrolled hiccup of the id, while he tries to ignore it and continue with his thought. But the OCD patient *becomes* the disease. If confronted with his behavioral patterns, an obsessive could not distance himself from the incessant symptoms by saying, "I can't stop myself from washing for some reason." He would embrace them, saying, "I'm always dirty."

This is not a disease in which the person suffers because she can't fight these destructive patterns that she recognizes as irrational. At its most extreme, when the patient has minimal insight into her disorder, she suffers because the patterns seem to be so rational, so imperative, so urgent, that she feels herself drowning in the catastrophic consequences of failing to carry them out. It is the endless array of rules and prohibitions and demands needed to hold back the floodwaters of dis-ease and disquiet.

As exemplified by the way that the American Psychiatric Association classifies OCD, this nightmarish disease is above all else a battle against anxiety, anxiety that can never be held off, no matter how much you count or check or wash. There can be brief instances of relief—when the street corner is reached during a walk and the number of steps turned out to be divisible by five, after all, there is a transient burst of comfort and safety: things are under control; things really will work out all right. But inevitably, the nameless dread returns.

The psychiatrist Judith Rapoport, author of the superb compendium of OCD case histories, *The Boy Who Couldn't Stop Washing,* has theorized about the ethological roots of OCD. She draws parallels between OCD and fixed action patterns in animals, small stereotyped behaviors that are carried out mechanically. We all know one familiar example of a fixed action pattern, the involuntary circling of a dog on its blanket before being able to settle down for the night. For a human, OCD is like a dog circling, but a dog, exhausted and bewildered, who can never, ever stop.

Remarkably, the disorder is coming to be understood on a biological level. It is responsive in some cases to certain neuroactive drugs. The brains of OCD sufferers show enhanced oxygen and glucose utilization in pathways related to motoric patterns—seemingly, there is a neurological itch, a metabolic drive in those brain regions toward fueling the arranging and washing and repetitive gesturing. Parts of those brain regions are even a different size from those in unaffected control subjects. There is even coming to be recognized a genetic component to the disorder as well.

The far-from-original point I will try to make now should seem obvious: There is an unsettling similarity between the rituals of the obsessive-compulsive and the rituals of the observantly religious. This parallelism might initially be difficult for many of us to appreciate. For those readers who are religious, it is often a rather liberalized version, in which intellectual or emotive components, or the carrying out of good deeds, might take precedent over a complex jungle of daily rules and rituals. But for the practitioners of most religious orthodoxies, rituals of cleansing of self and of food, of checking, of leaving and entering places, and of numerology consume each day.

To begin to make this point, I will borrow a trick used by many cultural anthropologists when trying to emphasize something about societies as a whole, which is to start with examples from a more alien culture, one that can likely be observed from a more detached, objective distance. For most readers, classical Hinduism fits that case.

For the observant Brahman, life is filled with a staggering, near-consuming array of rituals. The day's initial cleansing and prayers must begin two hours before sunrise, complete with instructions as to what one's first sight must be (something auspicious such as a ring, a coin, or a sacred cow) and which foot may first touch the ground (the right one). Defecation must be carried out in an open place, during which time the Brahman cannot look at the moon, at trees, at a tilled field, a temple, or even an anthill. The task completed, the washing begins—ten times of the left hand, seven times of the right hand, five times for both, followed by three washings of the feet and twelve unswallowed mouthfuls of water. More and more details follow involving bathing in the river, the proper repositioning of the loincloth, the seated position to be taken for the first prayer (which comes with complex instructions concerning pressing specified fingers against one of the nostrils and taking a certain number of breaths coinciding with recitation of one prayer, followed by pressing of the other nostril and another prayer). Hours' worth of rituals to carry out before even the first meal, everything carefully specified, with the prohibition that if something goes amiss, it must begin all over again.

And so goes the rest of the day. Prayers that must be recited in multiples of the number of beads on a holy bead ring, rules about the number of mouthfuls of food to be taken with a designated hand, multiples of times that a sacred plant must be circled when it is encountered, rules about not touching one's foot on the threshold of a temple when it is entered, the stone image of a bull at the doorway of the inner shrine that must be touched just so. The prescriptions go on and on.

Life for the Orthodox Jew is much the same. The rules of everyday life, the Halakha, are equally encompassing. The kosher laws are complex strictures regarding the cleansing and consuming of food—what species may be eaten, how animals are to be slaughtered and their meat prepared, the number of hours that must pass between a meat meal and a dairy meal. There are rules for the manner of hand washing, the volume of the ritual wine cup at a meal, the number of minutes before sunset that candles are lit to usher in the Sabbath. There are magical numbers—the number 18, which signifies life, the number 39, signifying the omnipotence of the Lord— that dictate the frequency of recitations of prayers, the number of times the threads of a prayer shawl are wound on themselves. Much as for the Brahman, the Orthodox Jew cannot enter sacred places without touching a holy icon, in this case a small casement containing a sanctified parchment, nailed to the doorway. And if a mistake is made, once again, there are complex requirements for making things right—for example, if utensils are contaminated by touching prohibited foods, they must be buried in soil for a certain length of time to purify them.

Among the most fascinating manifestations of the obsessiveness of Orthodox Judaism are the 613 rules that must be observed each and every day, 365 concerning things that must be avoided, 248 that must be carried out (a preponderance of things prohibited that led one medieval rabbi, perhaps of a depressive bent, to remark that we would have been better off if we had not been born). There is something remarkable about the rules that transcends the details of what they regulate. One might guess that over the years, the 613 rules accumulated, based on rabbinic decisions about proper ritual. Instead, in a fit of numerologic frenzy, the numbers came before the rules. Someone back when decided that there had to be 613 rules about daily life—365 prohibitions for the days of the year, and 248 required daily rituals corresponding to what was then thought to be the number of bones in the body—*without yet knowing what the rules were*. And rabbinic scholars have labored ever since to come up with the list, debating and differing with Talmudic fervor. Sometimes magic numbers emerge in religions as an aid to memory—it is not surprising that ten commandments, rather than nine or eleven, evolved among a nonliterate, nomadic tribal people who already used a base-ten numerical system. But there is no way that the 613 and its two parts are a memory device, to aid people in remembering the myriad (and still-debated) rules. The magicality of the numbers themselves is what resonates—each day, there are as many things that God requires of us as there are bones in the body, and as many things forbidden as there are days in a year. You can't ask for a more explicit case of numerologic ritual coming before content.

Everyday life is much the same for the classically observant Muslim. The foot with which one takes the first step of the day, the step into and out of the lavatory, are all specified. The number of times water is taken into the mouth for cleaning is specified, as is the washing of each hand and the sequence in which it is done. Washing of the rest of the body follows similarly, with the usual rules about returning to Go if there is a mistake—for example, if a man inadvertently touches his penis after he is otherwise clean, the process must begin again.

Prayers occur a certain number of times a day, facing in a certain direction. Complex prohibitions occur as to which animals can be slaughtered for food and how it is done. There are, once again, magical numbers—7, 10, 70, and 100 are ripe with significance. There are rigidly quantitative aspects to the numerology: a prayer said with clean teeth is 70 times better than one said with unclean teeth; public worship is 17 times better than private worship. So goes the day, ending with curling up for bed (lying on the right side specified) and facing Mecca.

The similarly obsessive qualities of Christian religions should now be apparent. Whether the numerology of the numbers of Hail Marys, the ritualistic use of rosaries, the details of baptism, the rules for entering and leaving churches, the magic invested in the number 3 or the number of Stations of the Cross, Roman Catholicism readily fits the patterns already discussed. The same is seen with the Protestant religions. Even among Lutherans (chosen for their relative lack of ornateness), there is no shortage of detailed rules—prayers differing as to time of day, prayers differing as to even- and odd-numbered years, precise orderings of baptism and of confession, rituals of foot washing in imitation of the washing of Christ's feet. In contrast to some of the other religions just discussed, there are also rules concerning the use of colors and of sound as part of daily ritual. For example, Lutheranism is full of precise strictures about music. The hymnal companion to the *Lutheran Book of Worship* contains a stern checklist of dos and don'ts for the organist, beginning with, of course, the requirement that the correct notes be played, but including strictures against allowing crisp dotted rhythms to slide into decadent triplet rhythms, and warnings to attack all the notes of a vertical chord simultaneously.

One could go on, but it is not necessary; it should be clear how the same patterns come up, again and again. The links between obsessive-compulsive rituals and religious rituals have long been noted. Naturally, Freud had something insightful to say about it. In his 1907 essay "Obsessive Acts and Religious Practices," he explicitly linked the two. In a marvelous sentence, he describes obsessional neurosis "as an individual religiosity and religion as a universal obsessional neurosis." In a similar vein, the psychoanalyst Robert Paul has more recently described religion as "the neurosis of civilization." In a 1975 essay, the psychiatrist Eugene d'Aquili and the anthropologist Charles Laughlin explicitly compared the fixed action patterns of animals with religious ritual.

Lest one think that this represents the secular mental health profession pathologizing religion, the link has long been recognized by theologians as well. In her book, Rapoport devotes an appendix to the arcane Catholic concept of "scrupulosity." As first discussed nearly five centuries ago by Ignatius Loyola and by scholars ever since, scrupulosity represents religious doubt and anxiety, manifesting itself in the pouring of energy into comforting ritual. Jeremy Taylor, a seventeenth-century theologian, gave an apt description: "They repent when they have not sinn'd. [Scruple] is a trouble where the trouble is over, a doubt when doubts are resolved." In a rough way, scrupulosity can be defined as the manifestations of obsessive anxiety within a Roman Catholic framework.

Surprisingly, these scholars often discuss scrupulosity in order to warn clergy about it. Priests have often been told to, in effect, be on the lookout for the practitioner who comes to their religion merely to quiet the frenzied imperative toward ritual. Priests are admonished to try to turn the sufferer toward the contemplation of the content of the rituals, the meaning of God, and their behaviors, rather than focusing on the sheer surface of the rituals themselves.

Similar admonitions are seen in other religions as well, in the face of practitioners who may become lost in the ritual for its own sake. In Orthodox Judaism, during the Sabbath reading of a chapter of the Torah, the reader must actually read the chapter, rather than recite it by memory. Two interpretations are traditionally given for this rule. One is practical—recitation by rote is more likely to contain errors. The other harks back to the same theme as Catholic scrupulosity—reading by memory or by rote promotes the feeling of empty ritual for its own sake. The Jewish tract the Ethics of the Fathers (2:18) warns, "Do not make your prayers routine." A similar rule occurs in Islam, harking back to Muhammad, who, in a surprisingly quantitative approach to the issue, said that a man receives credit only for the proportion of his prayers that he actually thinks about.

Thus, while the clergy often have to rein in the subset in whom a frenzy of ritual becomes unseemly, organized religion, and especially the organized religion of myriad daily rules and prohibitions and patterns, remains a remarkable haven for the obsessive-compulsive. Outside the context of religion, the obsessive-compulsive is, by definition of the disorder and its extent of disruption, a dysfunctional outsider, cut off from the community by the demands of the illness. Reframe those same behaviors in the context of religion and it all changes—the person is able to dwell at the structured core of a shared community. One might generate the prediction, untested to my knowledge, that obsessive-compulsives who are atheists or who are members of some of the sparer, more minimalist religions (the Quakers come to mind) will not fare as well as those sheltered by the more Byzantine religions.

The solace of religious ritual for the OCD patient is the

structure. Orthodox religion is a world of rules and strictures and requirements and prohibitions, a world of signposts. Again, the key to understanding OCD is its core of anxiety. And the key to understanding the solace that the obsessive-compulsive might find in religious ritual is not that it leads to a *diminishing* of that anxiety. That is not possible—not with our legacy of religious ritual as so often so febrile, so often the tinder for endless personal and societal immolations, not when the rules are so demanding and the punishments measured in eternity. It is an issue not of quantity of anxiety, but of quality. For the obsessive-compulsive, religious ritual provides the immeasurable comfort of the transition from nameless dread to dread that is abundantly named, measured, weighed, cataloged, and shared. It is a vast relief when the bogeyman comes with explicit instructions as to his care and feeding.

IV. The Emergence of the Godspoken

In *A Hunger Artist,* Kafka describes the travails of the protagonist, who is brought into the town square in a cage in order to carry out a forty-day fast for the edification of the populace. One initially assumes this to be one of Kafka's stranger parables, but there were, in fact, professional fasters of this sort throughout medieval Europe, the perfect entertainment for those perhaps bored with their season tickets to the bearbaiting. A remarkable thought—you could get paid to starve, something with a parallel in the present discussion. In many cultures, the advantages of intermixing obsessive-compulsive and religious rituals can even be taken a step further, beyond the realm of the lay practitioner. We live in the age where the religious leader and the psychotherapist often occupy roughly equivalent niches in their respective sectarian and nonsectarian ecosystems. We are rife with the images of such religious leaders as dispensing solace and insight—counseling the young couple before their marriage,

Mark Tansey, Doubting Thomas, *1986;*
private collection, Los Angeles, courtesy Curt Marcus Gallery, New York

comforting the mourner, ministering to our pains. But that is only one face of religious leadership. They can also be the purveyors of damnation, or the warriors at the front of columns of crusaders. And in many settings, it is the religious leaders who are simply the best at carrying out the rituals—the most fervent, energetic, or inventive. This is extraordinary—with any luck, *you might make a living being obsessive-compulsive.*

As before, this is often most readily appreciated by considering what is, for most readers, a more alien case. The life of the orthodox Hindu Brahman is consumed not only with his own arduous daily rituals, but with freelance work doing cleansing and numerological rituals for others. For example, the observant Hindu is expected to recite the Gayatri mantra 2,400,000 times in his lifetime. It is common for aging gentlemen, perhaps feeling the shadows lengthening and lacking the fortitude for repetition, to hire a Brahman to finish their homework for them, reciting the remaining mantras. Or for those in a hurry, one may hire 24 or, for the truly wealthy, 240 Brahmans to do all 2,400,000 mantras nonstop in one major blowout numerologyfest.

The examples closer to home are subtler and harder to recognize for their familiarity, but still represent the same. Consider the traditional Catholic practice of paying a stipend to a priest to have a Mass said for someone, or the endless circumstances in which a religious leader is brought in to ritualistically ensure a successful baseball game, high school commencement, or session of Congress.

This economic arrangement can even take on a quality of second-order derivatives, where a religious leader can be hired to ritualistically check that someone else is carrying out a ritual properly. For example, for a food to be labeled as kosher by the Union of Orthodox Rabbis, the factory or slaughterhouse where it is produced must hire a specially trained rabbi. It is not this individual's job to actually partake in the food preparation. Instead, his task is purely an observational one, assuring that the rituals of preparation of, for example, no-cholesterol tofu hot dogs, are carried out in a way that would bring an approving smile to the lips of the Patriarchs.

One may protest that the more conventional versions of being hired to perform religious rituals (i.e., the ones we are more familiar with in our own culture) do not really represent "hiring"—it is rare for a religious leader to charge for doing a graduation invocation, instead viewing it as part of the job. But that is precisely the point. It is the *job,* often full-time, with either the direct hiring by a community of a minister or rabbi, complete with contract and health insurance, or the indirect hiring of a priest through contributions to the Catholic Church. And in the cold terms of economic anthropology, the evolution of this sort of stratification requires the peasantry not only to earn their bread by the sweat of their brow, but to earn the bread for someone else as well, while the priestly class is busy with hand washing.

This idea is captured with an unexpected brilliance in what was to me an otherwise forgettable science fiction novel, Orson Scott Card's *Xenocide* (New York: Tor Books, 1991). In a multiplanetary and oppressive empire, there is a planet whose culture is dominated by a hereditary priestly caste, who are "godspoken." These religious leaders, intellectually gifted but useless, live sumptuously off the servile offerings of the peasants, passing their days in complex, exhausting rituals of purification and numerology. Late in the book, an insidious plot is revealed—this priestly caste was a genetic experiment of the evil overlords of the galactic empire. Long ago, they had recognized the highly developed intellect of the people of this planet and feared that, someday, they might prove to be able leaders of a revolution. As an effective preventative measure, the overlords introduced a genetically engineered virus that infected a subset of the population and transferred into them the gene for OCD. That subset, counting and washing and chanting, soon had evolved these rituals into religious ones, foisted them off onto the uninfected population at large with claims that their frantic imperatives were signs of being spoken to by god, and passed on the gene to their descendants. A parasitic priestly caste had emerged and the planet no longer constituted a revolutionary danger—the godspoken were too busy with the holy task of counting the number of lines in the planks of wood on their floors (a common compulsion among them), and the peasantry was too busy doing the cooking and laundry for them.

Thus, to the extent that religious ritual offers a sanctuary for the obsessive-compulsive, the most structured and rewarded of such safety is found in the priestly class. An extraordinarily modern, familiar cast to the suffering of an obsessive-compulsive is found in the case of a sixteenth-century Augustinian monk named Luder, whose writings have survived into our time. Anxious and neurasthenic, troubled with a relationship with a stern and demanding father, plagued with a variety of seemingly psychosomatic disorders, the young man had been caught one day in a frightening thunderstorm while walking alone, suffered a panic attack, and vowed to become a monk if he was allowed to survive.

True to his vow, he became a novitiate and threw himself into the rituals with a froth of repetition, self-doubt, and self-debasement. He described his dis-ease with the German word *Anfechtung,* which he defined as a sense of being utterly lost, a sense of anxious lack of mooring in every circumstance. He carried out each monkish ritual to perfection, urging himself to ever greater concern for detail, ever greater consciousness of God throughout the act, ever greater contrition for his own inadequacies . . . and would invariably find fault and have to start over again. The first Mass that he led was an agony of anxiety, as he was filled with fears of leaving out details, of saying something blasphemous. His spare hours of silent meditation were filled with obsessive, heretical thoughts, for which he confessed at length day after day. "I often repeated my confession and zealously performed my required penance," he wrote. "But I was always doubting and said, 'You did not perform that correctly. You were not contrite enough. You left that out of your confession.' " At one point, his father confessor, no doubt exhausted with having to hear hours of confessions each day from Luder, endless reportings of evidence of failings and God's justifiable anger, finally turned to the young monk with an exasperated, shockingly modern insight—"It is not God who is angry with you. It is you who is angry with God."

History gives us a final hint of this monk's affliction. He washed and washed, and it was all futile. "The more you cleanse yourself, the dirtier you get," he summarized plaintively. The vein of obsessive-compulsive anxiety is readily apparent in this young man, who would come to be known by the more modern version of his name, Martin Luther.

V. Who Invented Knocking on Wood?

Why should life for a certain type of religious individual and for the obsessive-compulsive be so filled with ritual? This has already been discussed in the context of OCD as an anxiety disorder, and even in the need the rest of us have to count stairs during troubled times. There is an inchoate comfort, security, even pleasure, that we tend to take in patterns and in repetition. D'Aquili and Laughlin proposed what they called a "biogenetic structuralism," speculating that our brains find there to be something intrinsically, electrophysiologically

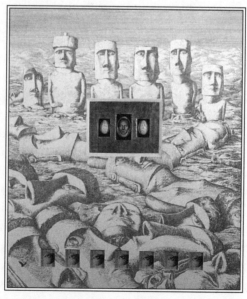

Scherer and Ouporov, The Birdman, 1995;
courtesy the artists

reinforcing about rhythmically repeated patterns. Their guess was that "repetitive auditory and visual stimuli can drive cortical rhythms and eventually produce an intensely pleasurable, ineffable experience in man." I know of virtually no neurobiological evidence for that idea, yet alone for why we get comfort not only from a ritual involving repetition over seconds (such as the repeated chanting of a mantra), in which there might actually be some sort of entrainment of cortical activity, but also from rituals with a cycle of repetition spanning a thousand Sabbaths. The biology of the comfort of repetition can't be quite that simple. Nevertheless, there is indeed a great and fundamental comfort that occurs when we encounter the familiar and repetitive, when we are reminded that we have trod this way before and that the ground is solid. It is a trait we share with many other species, a trait that becomes increasingly powerful in most of us as we age. (In Tracy Kidder's *Old Friends,* there is a wonderful comment by a nursing home resident, referring to one consequence of his roommate's quirks of aging, a comment to the effect of: To hear that story of Joe's once or twice is entertaining, to hear it a dozen times gets irritating, but to hear it every day for the last five years is one of the greatest things I can imagine.)

Thus, the need for ritual. But why should religion be filled not only with ritual, but so often with the same *types* of ritual— cleansing, food preparation, entering and leaving, numerology—as carried out by the OCD patient? For example, of the 613 rules concerning Orthodox Jewish life, most versions have approximately 100 of them concerned with food preparation. Why the similar foci of obsession in the two groups?

There have been some psychoanalytic speculations about this. There have also been some from biologically oriented anthropologists and psychologists. These individuals view ritual not merely as the random seizing upon some pattern, but as reflecting an underlying biological reality. In that view, it is no surprise that, in worlds full of communicable disease and potentially contaminated food, secular and nonsecular ritualism is highly concerned with issues of personal hygiene and of ensuring that food is clean and safe to eat.

These ideas are built around the notion that rituals of religion and of OCD are structurally similar because they arise from similar (psychoanalytic, ecological) roots. It strikes me that in many cases, there is a far simpler explanation, one that is, I believe, the only original idea in this piece. Religion does not just offer a refuge to OCD sufferers who learn to do the rituals that everyone else does, does not just particularly reward that with leadership positions among the individuals who excel in carrying out these rituals. It seems obvious—a lot of those rituals had to have been *started* by OCD individuals, their private attempts at structured anxiety reduction somehow turning into rules for everyone else. Insofar as a schizotypal individual, in the right place and time, might dramatically influence the very structure of a group's theological beliefs, an OCD sufferer, with a similar stumbling onto the stage of influence, might change the way a people carry out the rituals of their daily belief.

Who was the Hindu somewhere back when who had this obsessional thing about the number 24, the Muslim stuck on 70? Who was the Jew who believed that God had an inordinate fondness for the number of days of the year plus the number of bones in the body? Who invented worry beads, or wrapping the threads of a prayer shawl a set number of times? Who was the individual, in culture after culture, whose private, tortuous sequence of self-cleansing was the one that somehow caught on and became the standard version for the orthodox? In the vast majority of cases, we are never going to have a clue, these individuals and their needs lost in history.

There is something I find poignant about the idea that many religious rituals were once the private ones of individuals with OCD. There is a quality, noted by many clinicians, to the rituals of the OCD patient and that individual's relationship to them. The careful, exhausting, all-consuming frenzy of patterns that obsessive-compulsives are enslaved to often represent the most personal, most secretive corner of their lives. If their rituals are grounded in ones that are already part of their culture, it is often vital that they add something extra, something private, something more exhausting and challenging and demanding to keep them one step further ahead of the inevitable dis-ease, ahead of their *Anfechtung*. If their rituals have been invented out of the air, they are filled with a private symbolism, forming a template of the person's essence that can defy words, that can defy the patient's understanding of why these patterns seem so fundamental to their Self. The rituals may be shown in public at times, the sufferer may not be able to suppress them, but this is an intensely private act that is being displayed.

It is impossible to know in any given case how the transformation into the shared religious ritual occurs. During a time of trouble and community anxiety, perhaps others seize on the patterns displayed by this one influential individual, or perhaps she comes forward and offers it as a gift to the community—this is how I have pleased our God all these years. But the transformation occurs and that OCD individual is left with a yawning space of what is no longer her own. We are accustomed to the idea of the religious leader as sacrificing with a life of celibacy, abstinence, or poverty, with even life itself being sacrificed on a cross or in a conflagration. I suspect this transformation of ritual constitutes another case of such sacrifice: in the name of religious belief, in the name of concern in a time of trouble for the community to which one belongs, in the name of the good of the flock, the Secret is revealed, that private place of repetition and rule and pattern where, temporarily, the devil is held at bay and God knows you have done your best. It is given away to be franchised and distributed with what must seem like as much sanctity as an aerobic-exercise video. The ecosystem of belief provides some surprising niches of martyrdom.

VI. Epilogue: The Specter of Ersatz Faith and Ersatz Loss of Faith

I have gone on at great length to review these ideas that link the metamagical thought of the schizotypal to features of religious ideation, or that link the ritualistic imperatives of the obsessive-compulsive with features of religious ritualism. There are other, recent findings in neuropsychiatry that can be tossed into this arena as well.

Consider the case of "superstitious conditioning," a wonderfully apt phrase coined by B. F. Skinner in the early days of behaviorism. In many (although not all) realms, if you reward an organism for carrying out some behavior, you increase the

Alfredo Castañeda, Retrato del Combatiente Quintuple, *1989;*
courtesy Mary-Anne Martin/Fine Art, New York

likelihood of that organism behaving that way again. When done methodically, one can slowly "shape" an individual into carrying out an increasingly involved, complex task by progressively upping the ante as to what elicits a reward—there are endless examples of such conditioning, perhaps the most dramatic being the sort of training that goes into circus animals learning their tricks.

What if you give a reward at random intervals, irrespective of what the animal or person is doing? The response on the part of the individual, naturally, is to think something along the lines of "What did I just do to get that great reward? I must have done something." The individual repeats whatever behavior he was doing just before the reward came. He tries it again, tries a slight variant on it, prepares to try it another time, when . . . bingo, the next random reward shows up and his theory is resoundingly confirmed—"I knew it, do this behavior enough times and that *causes* the reward to happen." You've just conditioned a superstitious behavior.

Skinner showed this with his pigeons: reinforce some hungry pigeons with food at random intervals and you wind up with a bunch of animals exhibiting superstitious behaviors— in one cage will be a pigeon hopping vigorously on one leg, in the next, a pigeon alternately leaning from side to side, in another, a bird earnestly spinning in circles—each one convinced that *that* behavior is responsible for the intermittent reward. I've seen scientists conditioned into superstitious behavior plenty of times as well. Try some new difficult laboratory technique, one that hardly ever works. Inevitably, one day, out of nowhere, it works like a charm. In reality, it is due to some thoroughly obscure variable, one that the scientist can't even begin to guess at—the trace levels of some chemical in the water supply that day, or maybe sunspots, or perhaps fluctuations in the Tokyo stock exchange. But the scientist is so desperately pleased, so likely to leap on any and all random behaviors associated with this particular success, that all sorts of superstitious behaviors quickly get conditioned—since the scientist has no idea *which* behaviors were responsible for that unlikely success, they all get conditioned, virtually down to always wearing the same lucky pair of underwear on the day of the next experiment.

Do animals or people ever unlearn such superstitious behavior? In some cases, often when they are less dependent upon the reinforcing reward. It merely takes some variability, whether intentional or by accident, in the carrying out of the behavior—some details are missed, the behavior is done a bit wrong, the individual forgets to do it entirely, and . . . the reward still pops up. And that is about the time that the individual may begin to entertain that skeptical, agnostic thought, "Maybe it doesn't have to do with my wearing my lucky underwear after all."

Most broadly, what this teaches us is that during times of great need and dependency upon reinforcers, people (and other beasts as well) are particularly vulnerable to deciding that there are causes and effects going on that really aren't. And the leap from "superstition" as used by psychologists to its everyday use is obvious. A recent finding that anchors this in neurobiology: damage a certain part of the brain, and animals become more vulnerable to superstitious conditioning, less capable of discerning a cause-and-effect relationship in explaining how the world works. And out of this comes a corollary that is the typical next step for a neuroscientist—it could well be that individual differences in the workings of that part of the brain might explain individual differences in how readily superstitious behaviors occur.

There are other links between abnormalities of the brain and patterns of behavior relevant to this article. Neurologists are coming to recognize that certain types of epilepsies cause characteristic changes in the personalities of their sufferers, and not just around the time of a seizure. Perhaps the most fascinating are the changes seen in individuals with a certain type of seizure centered in the temporal lobes of the brain. As was reviewed in "How Big Is Yours?" these individuals evolve into people with a markedly serious, humorless style. They tend to be extremely rigid in their ways, shying away from novel situations and people, demonstrating a sticky or viscous personality. In addition, temporal lobe epileptics tend to become "hypergraphic," displaying a bizarre compulsion toward lengthy writing. Most interestingly, the disorder is also associated with developing a deep interest in religious and philosophical issues. This is not to say that this form of epilepsy is associated with becoming religious; rather, it appears to be associated with becoming interested in religion. (This distinction is shown in a wonderful story I heard about the late Norman Geschwind, the towering figure in neurology who first discerned temporal lobe personality. A medical resident on Geschwind's team was reporting on an adolescent with temporal lobe epilepsy he had just seen in the clinic. Perhaps wanting to poke the great man a bit as some sort of dominance display, the young resident announced that, oh, by the way, the patient didn't quite fit Geschwind's personality scheme. And why? Geschwind inquired. Well, he had asked the boy if he was religious, and he had said he wasn't. Geschwind led a troop of residents down the hall to find the patient, in order to display what he was sure this resident had missed. Geschwind turned to the epileptic and asked him if he was religious. No. And then he asked the critical follow-up: And why not? And the boy proceeded to give an impassioned half-hour lecture on Martin Buber's analysis of God's treatment of Job, on Bertrand Russell's critique of the Sermon on the Mount, on the Flood Story as myth as shown in the Epic of Gilgamesh . . . until Geschwind departed, satisfied.)

What does all of this mean, this intertwining of psychiatric and neurological disorders with faith? Despite the caveats I opened with at the beginning of this piece, some readers will be upset by the material covered. These are indeed disturbing ideas. But it is important to contain them by first emphasizing what disturbing things these ideas do NOT mean.

First, I am not saying that you've got to be crazy or neurologically impaired to be religious. Let's state that in a way slightly less in-your-face: It is not the case that any of a number of neuropsychiatric disorders are *prerequisites* for certain types of religious belief. Instead, the situation is much milder. For individuals with any of a number of neuropsychiatric disorders, certain types of religious beliefs or activities supply a remarkably good fit and a remarkably effective and consoling refuge.

Second, I am not saying anything quantitative, such as that most people, or even that a substantial minority of people, with certain religious orientations have come to them because of their neuropsychiatric difficulties. The quantitative issues are irrelevant; the points raised would be equally pressing if they applied to the religious beliefs of only a single person.

Finally comes the most subtle disclaimer. The purpose of this piece has not been to sarcastically or caustically emphasize that there are some people whose religious beliefs are nothing more than the mechanistic twitchings of their neuropsychiatric problems. The point is not that, for them, it is "just biological." I have no doubt that there are as many neuropsychiatric clues (although far less researched) as to why people lose faith as gain it and, as a fervent atheist, I find it equally fascinating to think about someone's atheism being "just biological" as would be someone's faith.

That is the main point. It does not matter in the slightest whether these biological insights tell us something about a million people or only about a single individual, or whether they tell us something about ardent faith or about the ardent loss of faith. It is the mechanistic implications of these studies that are, to me at least, most important. For many of us, how we feel about God, gods, or the absence of them, what moral imperatives we face here on earth and what rewards we might anticipate after-ward—the decisions we have made about our religious beliefs—are among the most intense, most personal, often most difficult, of our lives. They are among the stances that most define us as individuals. Half my adolescence was taken up with the con-sequences of my own struggles with those issues, and I still feel their echoes decades later; it is a rare week in which, at some point, I do not rage at this god for ceasing to exist for me, in which I do not lament the discrepancy between what I wish to feel and what I can. What does it mean if, amid this intensely personal journey of defining who we are, this often critical way in which we anchor our sense of self, there exists even a single person on this planet who has reached that state of resolution as a result of peculiar neurotransmitter levels?

This is the same disquieting point where we began this volume, a similar type of question that was posed toward the end of the first piece, "How Big Is Yours?" with its tour of frontal disinhibition syndromes, Tourette's, OCD, and other neuropsychiatric disorders. What will it mean as we learn about deterministic factors influencing what makes us who we are—how and whether we go about being religious, whether we take pleasure in novelty or in familiarity, how readily we speak our minds, whether our bent is toward monogamy or infidelity, and what the gender is of the person around whom these issues of faithfulness revolve? And the intervening pieces in this volume have simply added further to this question—what does it mean that there are biological constraints to how we move through our life histories, to how we go about some of our most personal junctures such as mourning? What do we make of the fact that at point after point, our traits prove us to be nothing more than a not-terribly-unique form of primate?

At the end of "How Big Is Yours?" I posed this question and tried to frame some of my own answers relevant to the societal or political implications of the biology of our behavior. What does it all mean? For one thing, that we must forever be on guard against the ideologues who would distort this science to design and justify the next holocaust. And it means that we must recognize that this science of what constrains and links us all into similar patterns should be a rallying point for our being more empathic, tolerant, and pluralistic, should be a call to recognize new biological realms in which people might need our help.

But as I've lectured to students about these subjects over the years, I've seen them become uncomfortable with the implications of this science at various points, and it is rarely because of their wrestling with the sociopolitical implications. Instead, it is the philosophical implications that plague and challenge our sense of who we are. What does it all mean?

It seems to me that people become upset for two related reasons. The first has to do with the sheer complexity of what biology is revealing to us—the stupefying array of neurotransmitters and synaptic connections, the intertwining of interactions between hormones and environment, the dizzying subtlety with which early experience can influence how our bodies work forever after. The problem here is not just that one can easily despair at ever understanding any of this. It's that this complexity runs counter to an image many of us have about living systems, where life and its processes are thought to unfold with a simplicity free of calculation or design. Natural. And instead, every facet seems the very antithesis of simplicity— where something even as seemingly straightforward as the timing of puberty involves our bodies machinating, ticking off the rules of evolution, monitoring the state of the world around us, adding and subtracting the costs and benefits of each step. All of this brings home a lesson learned by most biologists: To the extent that we think of "natural" as effortless, unplanned, unaffected, and simple, it is rarely the nature of our bodies or our behaviors to be natural. We cannot afford such a luxury.

The more resonant way in which people often become

upset at this point has to do with so many of these findings in behavioral biology running counter to an image many of us have about ourselves. There is a pleasure that most of us hold in our breasts, the notion that we are each a mysterious, shimmering flame of uniqueness. And there is the fear that that is precisely what the scientists will try to capture and thus destroy as a result of gaining this sort of knowledge—the image of the never-to-be-seen-again snowflake melting because of the warm breath of someone looking too closely, the image of a unicorn hemmed into a corral, gawked at by crowds oblivious to its sadness and loss.

This image is also encompassed in the classic Arthur C. Clarke short story "The Nine Billion Names of God." It is the perfect metaphor for the fear that many people have about where scientific knowledge will lead us and the price it will exact. In the story, Tibetan monks team up with some scientists who possess the most powerful high-speed computer on earth. The goal in their collaboration is to generate the nine billion different names of God posited by Tibetans to exist. And as they come close to obtaining this impossible, unobtainable knowledge, the inevitable happens: the stars in the sky begin to fade.

This is what many people fear, that scientists, as they learn more and more about less and less, will inadvertently go and explain everything, will turn us with their knowledge into sterile assemblages of molecules and networks and equations, and that with each new fact accrued, another light in the heavens will be extinguished.

I do not worry about this. This is not because I am an unemotional scientist just intent on reaching that holy grail of turning dancing flames into equations. I am a reasonably emotional person and that is central to the science that I do. Yet I am not worried if the scientists inadvertently go and explain everything. This is for a very simple reason: an impala sprinting across the savannah can be reduced to biomechanics, and Bach can be reduced to counterpoint, yet that does not decrease one iota our ability to shiver as we experience impalas leaping or Bach thundering. We can only gain and grow with each discovery that there is structure underlying the most accessible levels of things that fill us with awe.

But there is an even stronger reason why I am not afraid that scientists will inadvertently go and explain everything— it will never happen. While in certain realms, it may prove to be the case that science could explain *any*thing, it will never explain *every*thing. As should be obvious after all these pages, as part of the scientific process, for every question answered, a dozen newer ones are generated. And they are usually far more puzzling, more challenging than the prior problems. This was stated wonderfully in a quote by a geneticist named Haldane earlier in the century: "Life is not only stranger than we imagine, it is stranger than we can imagine." We will never have our flames extinguished by knowledge. The purpose of science is not to cure us of our sense of mystery and wonder, but to constantly reinvent and reinvigorate it.

FURTHER READING

Some references regarding schizotypal personality disorder were given at the end of the first essay, "How Big Is Yours?" An introduction to schizophrenia can be found in most psychology textbooks. However, for a moving, evocative view of the disease, read Susan Sheehan's *Is There No Place on Earth for Me?* (New York: Vintage Books, 1982).

The psychopathology of shamanism is discussed in a number of places. Paul Radin and his thinking are considered in S. Diamond, "Paul Radin," in S. Silverman, ed., *Totems and Teachers: Perspectives on the History of Anthropology* (New York: Columbia University Press, 1981). The quote by Radin, as well as by his Winnebago informant, came from this article. Also see E. Ackerknecht, "Psychopathology, Primitive Medicine and Primitive Culture," *Bulletin of the History of Medicine* 14 (1946): 30; J. Silverman, "Shamans and Acute Schizophrenia," *American Anthropologist* 69 (1967):

21; G. Devereux, "A Sociological Theory of Schizophrenia," *"Psychoanalytical Review* 26 (1939): 338; R. Shweder, "Aspects of Cognition in Zinacanteco Shamans: Experimental Results," in W. Lessa and E. Vogt, eds., *Reader in Comparative Religion* (New York: HarperCollins, 1979) 327; A. Kroeber, "Psychosis or Social Sanction?" *Character and Personality* 8 (1940): 204. For a historical perspective on the subject, also see J. Atkinson, "Shamanisms Today," *Annual Review of Anthropology* 21 (1992): 307.

The statistics regarding the supernatural beliefs of Americans come from G. Gallup and J. Castelli, *The People's Religion: American Faith in the Nineties* (New York: Macmillan, 1989); and G. Gallup and F. Newport, "Belief in Paranormal Phenomenon among Adult Americans," *Skeptical Inquirer* 15 (1991): 137.

The Beecher quote can be found in J. Braude, *Lifetime Speaker's Encyclopedia* (Englewood Cliffs, N.J.: Prentice Hall, 1962), 682.

As referenced at the end of "How Big Is Yours?" the most accessible book on obsessive-compulsive disorder is Judith Rapoport's *The Boy Who Couldn't Stop Washing* (New York: Signet, 1989). More technical discussions of the frequencies of different ritualistic behaviors in OCD patients can be found in S. Rasmussen and J. Eisen, "The Epidemiology and Differential Diagnosis of Obsessive-Compulsive Disorder," *Journal of Clinical Psychiatry* 53 (1992): 4; J. Rapoport, S. Swedo, and H. Leonard, "Childhood Obsessive-Compulsive Disorder," *Journal of Clinical Psychiatry* 53 (1992): 11; T. Insel, "Phenomenology of Obsessive-Compulsive Disorder," *Journal of Clinical Psychiatry* 51 (1990): 4. Some technical discussions of the possible neurobiological correlates of OCD can be found in D. Robinson, H. Wu, R. Munne, M. Ashtari, J. Ma, J. Alvir, G. Lerner, A. M. Koreen, K. Cole, and B. Bogerts, "Reduced Caudate Nucleus Volume in Obsessive-Compulsive Disorder," *Archives of General Psychiatry* 52 (1995): 393.

The ritualistic content of Hinduism can be found in M. S. Stevenson, *The Rites of the Twice-Born* (Munshiram Manoharlal, Delhi: Oriental Books Reprint Corp., 1971). Jewish rituals are found in H. Donin, *To Be a Jew* (New York: Basic Books, 1972). The 613 rules in Orthodox Judaism are discussed in Rabbi Dr. Chavel, trans., *Sefer Ha-Mitzvoth of Maimonides in Two Volumes* (London: The Soucino Press, 1967); and *Sefer ha Hinnuch: The Book of Mitzvah Education,* ascribed to Rabbi Aaron ha Levi of Barcelona, based on the first edition (Venice, 1523) (Jerusalem, N.Y.: Feldheim Publishers, 1989). Note that this represents a case of the two lists of the 613 rules differing.

The ritualism of Islam is apparent in Al-Ghazzali, *The Faith and Practice of Al-Ghazzali,* translated by W. Montgomery Watt (Chicago: Kazi, 1982). Lutheran ritual is covered in the Lutheran Church of America's *Lutheran Book of Worship,* Ministers Desk Edition (Minneapolis: Augsburg Publishing House, 1978); also see M. Stulken, *Hymnal Companion to the Lutheran Book of Worship* (Philadelphia: Fortress Press, 1981).

Freud's thought on the subject can be found in his 1907 work "Obsessive Acts and Religious Practices." *Collected Papers of Sigmund Freud* (New York: Collier Books, 1963). Also see R. Paul, "Freud's Anthropology: A Reading of the 'Cultural Books,' " in E. Garcia, ed., *Understanding Freud: The Man and His Ideas* (New York: New York University Press, 1992); and E. d'Aquili and C. Laughlin, "The Biopsychological Determinants of Religious Ritual Behavior," *Zygon* 10 (1975): 32.

Orson Scott Card's 1991 science fiction novel, *Xenocide,* is published by Tor Books, New York.

Luther has been considered in endless biographies. The classic one delving into his psychological makeup is Erik Erikson's *Young Man Luther* (New York: Norton Library, 1958), the work that defines psychobiography. Quotes by and about him come from this work.

An introduction to superstitious conditioning can be found in most psychology textbooks. One might also consult any of a number of books by Skinner on conditioning and learning, including B. F. Skinner, *About Behaviorism* (New York: Vintage, 1985).

References regarding temporal lobe personality disorder can be found at the end of the first essay in this book.

Arthur C. Clarke's "Nine Billion Names of God" can be found in a collection by the same name (New York: Ballantine Books, 1953). Versions of J. B. S. Haldane's quote appeared in numerous of his books, including *Adventures of a Biologist* (New York: Harper and Brothers, 1940).